T0227663

Latin American Society
Second Edition

Tessa Cubitt

LATIN AMERICAN SOCIETY
Second Edition

Routledge
Taylor & Francis Group

LONDON AND NEW YORK

First published 1988 by Pearson Education Limited
Second edition 1995

Published 2014 by Routledge
2 Park Square, Milton Park, Abingdon, Oxfordshire OX14 4RN
711 Third Avenue, New York, NY 10017, USA

First issued in hardback 2017

*Routledge is an imprint of the Taylor & Francis Group,
an informa business*

British Library Cataloguing in Publication Data
A catalogue entry for this title is available from the British Library.

Library of Congress Cataloging-in-Publication Data
Cubitt, Tessa, 1944-
    Latin American society/Tessa Cubitt.—2nd ed.
    p.   cm.
    Includes bibliographical references and index.
    ISBN 0-470-23469-5
    1. Latin America–Social conditions—1982- 2. Economic
development–Social aspects.   I. Title.
HN110.5.A8C83   1995
306'.098–dc20   94-24810
                                                          CIP

Set by 8 in 10/11pt Ehrhardt

ISBN 13: 978-1-138-17980-6 (hbk)
ISBN 13: 978-0-582-22548-0 (pbk)

For David and Christian

Contents

List of figures and tables

Acknowledgements

I am especially indebted to Michael Redclift for all his encouragement in getting the project off the ground and the invaluable assistance he gave me over the text. For their helpful comments on the original script, I would also like to thank Caroline Ramazanoglu, David Corkill, Sue Cunningham and David Lehmann. The fact that the second edition got to the printers at all is largely due to the help, research and resourcefulness of Juanita Underwood. I am particularly grateful to Sundra Winterbotham for her thoroughness in typing the manuscript and all the trouble to which she went to get it done on time. Special thanks are due to David, my husband, for all his support and help with sources and style. Finally I want to thank my students in Portsmouth, who over the years have made my Latin American classes a challenging testing ground of ideas and theories.

I am grateful to the following for permission to reproduce copyright material: Latin American Research Review for Table 8.2 (Portes, 1985); Longman Group Ltd. for Fig. 1.1 (Todaro, 1992); United Nations for Table 5.2 (Cecillia López and Molly Pollack, The incorporation of women in development policies, *CEPAL Review*, **39** (LC/G. 1583-P), Santiago, Chile, December 1989, Table 5), Table 5.4 (Irma Arriagada, Unequal participation by women in the working world, *CEPAL Review*, **40** (LC/G. 1613-P), Santiago, Chile, April 1990, Table 1) and Table 7.5 (Norberto Garcia and Victor Tokman, Changes in employment and the crisis, *CEPAL Review*, **24** (LC/G. 1324), Santiago, Chile, December 1984, Table 2, United Nations publication, Sales No. E.84.II.G.5).

Whilst every effort has been made to trace the owners of copyright material, in a few cases this has proved impossible and we take this opportunity to offer our apologies to any copyright holders whose rights we may have unwittingly infringed.

Tessa Cubitt
February 1994

Introduction

This book is intended to be introductory. As it covers the very broad topic of Latin American society the aim is to acquaint the reader with some of the major issues and debates concerning Latin American society, offering references which can be used to follow up points in more detail if desired. The chapters, therefore, relate to distinct social areas, but there are also clear themes linking the sections of the work and providing a general, overall picture. Although the principal perspective is sociological, the material and theory used are taken from a variety of social science disciplines and the volume is therefore intended as a starting point for any study of Latin America within the context of development. I am concerned to discuss Latin America not in isolation, but very much in relation to the Third World in general, and to relate its social organisation to general development problems.

The first two chapters discuss development in general, firstly the meaning and implications of the term and then evaluating the theories available for analysis. At the end of the first chapter, some general points are made about Latin America in relation to other less developed regions. Development is defined in terms of movement towards a complex set of welfare goals, and the extent to which Latin America is achieving these goals, such as reduction in poverty, inequality and unemployment, and improved standards of education is referred to throughout the various chapters on the different sectors of society.

Chapters 3–9 turn specifically to Latin America, without losing sight of the broader implications. One of the themes of the volume is that despite very impressive economic growth rates in some areas and during certain periods in the last three decades, little progress has been made in the achievement of certain welfare goals, such as the reduction of poverty and underemployment, and in one area in particular, inequality, the situation if anything has worsened. It is argued that the benefits of development – which are very real – are only enjoyed by a minority, who maintain a very high standard of living, while the gap between the wealthy and the poorest in society becomes wider.

Chapters 3 and 4 demonstrate an increasing integration of society, as ethnicity gives way to class, when peripheral groups become more involved in the development process at times through political actions in defence of rights to land, and then as a universal value system becomes more widely accepted through the expansion of education and the spread of mass media. This does little to reduce inequalities and often only serves to heighten awareness of them. Chapter 6 also shows an increasing integration, this time of the peasantry, into the national economic system, but again the benefits are only felt by a few, the majority suffering from underemployment since most end up working in the informal sector (Chapter 7). For some, this presents opportunities; for the majority, although conditions are poor, poverty is not as great as in rural areas.

Chapter 8 provides an overall picture of Latin American society, in terms of class structure, showing the general situation which has been brought about by the changes described in the previous chapters. Combining class analysis and income distribution figures reveals the way in which social inequality has become entrenched. The middle classes have promoted economic growth, which has given them a style of life similar in some ways to that of industrialised countries, but have been reluctant to see any radical reordering of society. The role of the state towards development is discussed in Chapter 9, with attention being paid firstly to the authoritarian state, whose policies exacerbated the growth-without-distribution pattern, and then to the neo-liberalism of the democracies that emerged in the 1980s.

Despite the continuation of some problems associated with underdevelopment, Latin America is very much a dynamic society and emphasis in this volume is placed on the positive action of different sectors of society. The high levels of production are themselves an indication of the economic dynamism. The response of other sectors to problems has also been positive. The peasantry migrate long distances in search of work, they have responded to economic opportunity where this has presented itself and they have at times formed political movements. Many of the urban poor put up their own homes because standard housing is not available, and create economic activities from very little. Trade unions have brought about some advances for the working class. Recently this dynamism has manifested itself in the growth of social movements which cover a wide social area – urban movements, rural groups, gender-based associations, church groups – which, though they have all so far been limited in their achievements, seem to be gaining in strength and scope.

Latin America is by no means static, but the changes taking place bring mixed results in terms of development goals. The conclusion discusses the industrialisation models that so many Latin American countries have followed and the implications for other Third World countries.

Development

The year 1985, the time of writing this book, marked a high point in relations between the poor and the rich countries of the world, whereas 1994, the year of this update, has marked a low. Why? In 1985 the consciousness of the horrors of starvation and poverty were aroused universally by the tireless efforts of an Irish pop singer. It was the music of the bands and the involvement of international stars that made people aware of the destitution that some fellow citizens of the world experienced. In 1994 image after image of this destitution still continues to flow from our television screens.

What 1984–85 was to Sahel Africa, 1992–93 has been for the Horn of Africa. Disasters in Somalia, Mozambique and Ethiopia have witnessed suffering and starvation on a massive scale as in Upper Volta and Chad in previous years. Futile emergency aid continues to be the only possible measure introduced to save people's lives. Hope for a decade of development in the 1990s and constructive involvement of the United Nations in the world's problems has dissipated in a deepening world recession. Many lives and hopes have been shattered in the face of continuing external debt, faltering economic growth, increasing unemployment, growing civil strife, rising ethnic tensions, threats to the environment and the persistence of abject poverty. The UN has seemingly proved ineffective in combating these human causes of the starvation of the people of many Third World regions. In 1990 over 100 million people were effected by famine (Human Development Report (HDR) 1992: 14).

At the same time, in most industrialised parts of the world, there are increasing numbers of people who suffer from the opposite condition –

an excess of food and drink. Obesity and an increasing incidence of disease connected with an over-indulgence of rich and fatty foods and alcohol have led to a boom in the slimming business.

While one-half of the world's population is desperately short of calories, others are searching for the low-calorie version of foodstuffs and drinks. Supermarket shelves abound with slimming aids, and health farms attend to those for whom the problem has reached a more advanced stage. The affluence of the industrialised world and its resulting comfortable life-style has also led to greater concern for fitness in general, so that jogging has become an accepted part of contemporary life in Europe and the USA.

Are these stark contrasts inevitable, or is it possible to improve the living standards of the poor countries of the world? If the latter is the case, and the potential to reduce their problems exists, why has this not already happened? This book aims to examine these issues in an attempt to understand how and why they have come about and what solutions are available. The book, therefore, concentrates on the notion of development which, although often used very differently by various writers, is usually seen as the means to improved living conditions. Development issues are discussed in general, though the book focuses on Latin America as the particular unit of study.

WHAT IS DEVELOPMENT?

'Development never will be, and never can be, defined to universal satisfaction. It refers, broadly speaking, to desirable social and economic progress, and people will always have different views about what is desirable' (Brandt 1980: 48). This comment from the Brandt Report sums up the essence of the controversy in the notion of development. It must, by definition, mean progress, but this implies a value judgement, and what constitutes an improvement varies considerably according to views and ideology. Having defined development as 'a process of improvement with respect to a set of values' Colman and Nixon come to the conclusion that 'the rate or the relative level of a country's development are normative concepts whose definition and measurement depend upon the value judgements of the analysts involved' (Colman and Nixon 1986: 2). Colman and Nixon also point out what is widely supported by social scientists today, that 'value judgements are inescapable elements of factual study in the social sciences' (Colman and Nixon 1986: 2). This problem should by no means be a stumbling block to further analysis, but should serve to keep us alert to ideological frameworks within which judgements are made. Moreover, despite different viewpoints, there is often considerable agreement over the major goals of development. The objectives for development identified by Seers (in Baster 1972), for instance, would probably meet with wide-spread acceptance. These are:

1. family incomes should be sufficient to provide food, shelter, clothing and footwear at subsistence level;
2. jobs should be available to all family heads;
3. education and literacy rates should be raised;
4. the people should be given the opportunity to participate in government;
5. national independence is an important aim for each country.

These objectives cover economic factors necessary for survival, cultural factors relating to the satisfaction of non-material needs, and political factors allowing for the greater involvement of people in the running of their own affairs.

Attempts have been made to escape from an ideological standpoint in defining this area. Norman Long, in his *Introduction to the sociology of rural development*, refers to the focus of his book in the following words: 'my basic subject matter is sociological work that deals with the process by which rural populations of the Third World are drawn into the wider national and international economy and with the accompanying social transformation and local level responses' (Long 1977: 4). Here, attention is centred on processes and social changes of a particular nature, that is, involving of people in the wider society without any references as to whether or not this can be judged 'better' for them.

Social and economic progress refers to a very wide arena, the component parts of which may not all be progressing at the same rate. Can the changes that have been taking place in Brazil in the last few decades be called development? Transnationals, using advanced technology, operate in Sao Paulo, producing cars for the wealthy; modern buildings tower above the cities, and the luxury goods on sale are a testimony to an affluence that exists in some sectors of society. In the same country, however, live people who have some of the most simple forms of technology to be found in the world. The Amazonian Indians, many of them living in hunting–gathering bands, exist today in the same way as they have done for centuries. Since some of these tribes have had no contact at all with outsiders, their existence is only known to us through aerial pictures showing their dwellings in clearings in the jungle. Tribes such as the Nambikwara live on fruit, honey, roots, spiders, grass-hoppers, snakes and lizards, which the women collect, and game if the men are successful with their hunting (Brain 1972). The implements needed for this subsistence economy are minimal. How does one, then, categorise a country harbouring such extremes in terms of development? Development for whom? is a critical question that arises. Whether or not it is possible to bring about improvements for all is highly debated but, for most people, the concept of development means better standards of living for at least a majority of people, if not the mass of society.

Implicit in the term is an economic element and, indeed many would argue that economic changes are necessary to bring about other desired aims. It is important, however, to distinguish at this early stage between economic growth, which usually refers to some quantifiable index such as increase in per capita income, and economic development, which involves some kind of structural and organisational transformation of society. Thus, development does not just mean a little more of everything, it also includes a rather deeper change in societal arrangements.

In order to examine further all these issues in the Third World, it is worth taking a fairly general definition, which does imply a value-judgement but bears in mind the implications of this. Brookfield says that 'the popular trend is to define development in terms of progress towards a complex of welfare goals, such as reduction of poverty and unemployment and diminution of inequality' (Brookfield 1975: xi). How far do the changes that have been taking place in the Third World bring the countries concerned closer to these goals?

The whole essence of progress is change and, in the case of the Third World, the main changes that have been taking place are linked to the growth of modern society, in most cases based on the development of capitalism, though in a very small number of countries, notably China, Cuba and North Vietnam, revolution has brought about a move to a socialist society. Therefore, for the vast majority of the Third World's population, it is the growth of modern capitalist society which brings about alterations in their lives. It is the contention of this book that these processes, which draw different social groups into national society, have brought about both improvements and major problems. This uneven development is a major characteristic of non-industrialised countries. In the case of Latin America, uneven development is reflected in an increasing polarisation of society. Improvements have been concentrated among the 'haves' while the problems for the poor, far from being alleviated, become more onerous. In attempting to understand this, it is first necessary to clarify the multiplicity of terminology that has been applied to the part of the world under discussion.

Interest in the Third World was aroused in a meaningful way in the post-war period, when nations came together to discuss the future of world development. The non-industrial countries were then referred to as 'underdeveloped' in a famous UN report of 1951, which attempted to produce measures for economic development in the Third World. Later, however, the term 'underdeveloped' came to be thought of as derogatory, as if it implied that lack of development was due to a weakness or a failure within these countries. Emphasis shifted from 'underdeveloped' to 'developing', which had a more positive ring about it since it suggests that, although these countries may not be so economically advanced as the USA or Europe, this is not so much due to intrinsic inability as lagging progress, and the implication is that movement forward is taking place. 'Developing' held sway for some time until what may seem a rather obvious criticism gained considerable ground, and that is that many of these countries are clearly, not advancing and they have been unable to produce sustained, autonomous economic growth. The rather more subtle term of 'less developed country' (LDC) then appeared and has since become very widely used. Many development economists refer to LDCs when talking about this part of the world, since the term points out the disparity between the industrial and the non-industrialised world but, at the same time, tries to avoid value-loaded expressions. The terminology discussion has now gone full circle, because 'underdeveloped' has become the appropriate term for radical scholars who view the problems of the Third World as a direct result of the development of the industrialised countries. However, many of the development reports of the 1990s have continued to use the term 'developing'.

Attempts have been made to escape all the problems of value-judgements, relativity and the direction of processes, by using terminology that includes no reference at all to development. The Brandt Report chose to depict these areas in the geographical terms of North and South, pointing out that the vast majority of LDCs are in the southern hemisphere, while the industrialised societies predominate in the North (Brandt 1980). Although this terminology is attractively neutral in value terms, it does entail other difficulties, notably that it is not precise. Australia and New Zealand are very definitely southern countries, but not normally considered within the rubric of 'less developed'. Since the term LDCs and 'developing countries' have achieved such common currency, these are the terms used in this book.

CENTRAL ISSUES – POVERTY AND RELATED PROBLEMS

Thus the question that needs to be answered is, have the changes in the LDCs brought about progress towards the complex of welfare goals mentioned earlier? In order to answer this we have to decide first which welfare goals are important, and then if and how progress in these can be measured.

One of the obvious welfare goals which would be widely accepted is the alleviation of poverty. In the World Bank World Development Report it was estimated that there were a billion (1133 million) people in the world living in abject poverty, suggesting that one-fifth of the world's population are only just surviving. (World Bank 1992: 25, 29.) Though there is poverty in developed nations it bears no comparison with the deprivation experienced in the LDCs. What abject or 'absolute' (the term most commonly used by social scientists) poverty means is living life totally preoccupied with day-to-day survival. For the destitute there is very little work available and what there is, is extremely poorly paid. The major business of the day is finding sufficient food to keep a family alive. Housing is usually constructed by the poor themselves out of any available materials, including rubbish, and since these dwellings are not standard housing they are invariably not serviced by urban facilities such as roads, electricity, water supplies and sewage systems. Life is short for all: not only do people die young, but families can expect to lose one child in infancy. Diseases that could be cured in developed countries prove fatal because medicines are just too expensive.

The problems of poverty are expressed in a number of ways. Hunger and starvation are, perhaps, the most acute expressions of extreme poverty. Susan George graphically points out that, in the time it takes to read her book (about six hours) 'somewhere in the world 2500 people will have died of starvation or of hunger-related illnesses' (George 1976: 19). In the developing world 35 000 children under the age of five suffer avoidable death daily (World Resources Institute 1992: 82). Certain regions of the Third World experience these problems particularly forcefully. In the north-east of Brazil, most children who do not survive the first twelve months of life, die from hunger or illnesses associated with malnutrition. The problem is not one of shortage of food, but of distribution, for the resources and technical skills are available in the world today to feed the population of our planet (George 1976; Redclift 1984b).

Thus hunger persists because poor people cannot afford to buy food or produce it, rather than because supplies are inadequate in any absolute sense. Although the wealthier nations do eat considerably more than the poor, this is only part of the problem. According to the Third World Guide (1992: 86): 'The number of people who could adequately be nourished from the amount of land, water and energy freed from growing grain and soya beans to feed US livestock if North Americans reduced their intake of meat by 10 per cent is 60 million people.' Part of the problem, therefore, is that grain that could be used to nourish hungry populations of the poor countries is being used to feed livestock which, in turn will provide meat for the wealthy nations. In the cities of Latin America, the wealthy can obtain luxury imported foods, while hunger reigns in the countryside.

Malnutrition has various long-term effects on a population. Children are often the most vulnerable. It is, perhaps, at its most crucial in the first few weeks of life, for babies who are malnourished in their first few months may feel the effects throughout their lives. Since one child in six in LDCs (25 million children) is severely undernourished (World

Resources Institute 1992: 84) its influence on the adult population is extensive. In the rural highland areas of Peru, 75 per cent of children under five are malnourished. Eighty out of a thousand die before their first birthday (Poole and Renique 1992: 23).

Malnutrition effects both mental and physical development. Experiments have shown that adults who were undernourished as babies have lower IQs than those who have had sufficient calories, often to the extent of not being able to hold productive jobs (George 1976).

Apart from hunger-related illnesses, so often given visual coverage by the media in the form of pathetic children with swollen stomachs, undernourishment leads to physical weaknesses and a greater vulnerability to disease in general. Illnesses which most children expect to get and recover from in industrialised societies prove fatal among the poor. Perhaps the most outrageous statistic is that for Guatemala in 1981 – the last year for which figures are available – more than 70 per cent of deaths of children between the age of one and five were from easily preventable diseases (Painter 1987: 4). 'Malnutrition is probably the biggest single contributor to childhood mortality in underdeveloped countries' (Murdoch 1982: 99). 'Of roughly 37 million people who died in developing countries in the mid-1980s almost 37 per cent were children. This shocking figure compares to developed countries where only 3 per cent of annual deaths are children' (World Resources Institute 1992: 75).

Since malnutrition is not due to inadequate supplies, this is a problem which, in theory, could be solved. However, hunger has always been present in the world and as pointed out, is still widespread. Thus, an important aspect of development must be a better-fed population. Attempts to bring this about, in terms of the Green Revolution and food aid projects, will be examined in Chapter 6 on rural society.

Apart from malnutrition, people in LDCs suffer from poor health in many other ways. The population of Sierra Leone can expect to live a shorter life than anywhere else in the world, with a life expectancy of forty-three years. The average person in Guinea and the Gambia is not that much better off, with a life expectancy of forty-five, and in Angola, the Niger, Ethiopia and Somalia the population can expect to live to forty-seven (World Resources Institute 1992: 249). This is less startling when one realises that 1.5 billion people lack access to health services globally (HDR 1992: 14). In the highland departments of Apurímac and Huancavelica, in Peru, there are only 0.4 doctors for every 10 000 people (Poole and Renique 1992: 24).

As well as the lack of adequate food, poverty means the inability to acquire the other basics of life – clothing, shelter, access to health facilities and education, and sources of mental and spiritual sustenance. What makes poverty particularly glaring is the existence, so often, of wealthy groups within the same country. Gross inequality within nations is very much a characteristic of the underdeveloped world in general, though more pronounced in Latin America than elsewhere.

Absolute poverty is not just shown up by these contrasts, but is also closely related to this issue which is, in itself, another problem of under-development, that of inequality. Todaro argues that 'the magnitude of absolute poverty results from a combination of low per capita incomes and highly unequal distributions of that income' (Todaro 1992: 158). There are, therefore, the twin problems of the low level of total earnings of a nation and the way these are distributed amongst its members. At first sight, it may seem that they could be solved by increasing a society's Gross National Product (GNP), that is, the sum of the nation's wealth. If this were the case, then attention could be focused on increasing productivity, assuming that the new wealth would 'trickle down' through society, enriching the population in general. However, there is now considerable evidence to show that increased productivity does not necessarily benefit the mass of the population and, therefore, it is not enough just to attempt to raise GNP and hope that a redistribution will follow, or even that the majority will experience some improvement. What is more important than the increase in production, is the character of economic growth. This refers to the way in which growth is achieved, which groups in society participate, which sectors are given priority and which institutional arrangements are emphasised.

Todaro offers some useful evidence for this. He has examined the relationship between economic growth, based on increased GNP and the distribution of income, in Third World countries. Taking thirteen developing countries, he plots on a graph their rates of growth of GNP on the horizontal axis and the growth rate of increase of the poorest 40 per cent of the population on the vertical axis. Each country's data, calculated for a period of six to seven years, are plotted on the graph at a point which reflects the intersection of the two variables concerned. The countries above the 45° line have experienced an improvement in distribution of income, i.e. the incomes of the poorest 40 per cent grew more rapidly than the overall GNP rates, while countries below the 45° line are those where distribution has worsened.

What the diagram reveals is that there is no obvious relationship between GNP growth and distribution of income. Countries are fairly evenly scattered above and below the line, showing that increased GNP does not of necessity, lead either to an improvement or a deterioration of income distribution. (See Fig. 1.1.)

Another study, which was wider-ranging, analysed the relationship between the shares of income accruing to the poorest 60 per cent of the population and the country's economic performance, in forty-three developing nations. The main conclusion that can be drawn from this is that the impact of economic development on income distribution has been in general to decrease both the absolute and the relative incomes of the poor. Todaro suggests that growth tends to lead to a 'trickle-up' process rather than the reverse (Todaro 1992: 161).

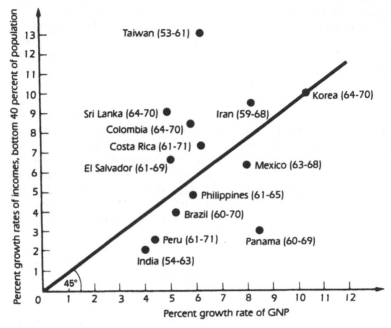

Fig. 1.1   Growth, poverty and income distribution. (Todaro, 1992: 160)

There are some economists who, following the pioneering work of Professor Kuznets, suggest that in the early stages of economic growth, the gap between the rich and the poor becomes wider, but will narrow in later stages. Others develop this point by arguing that a widening of the distribution of income levels is not necessarily economically undesirable in the long run because it provides a basis for further economic growth. This view is supported by the argument that inequality is necessary to provide savings for investment of a sufficient magnitude to promote economic growth. It is necessary to have some concentration of wealth available for capital investment.

Clayton argues that a preoccupation with the redistribution of wealth has distracted academics from the important issues of development. He suggests that this attitude is economically harmful and politically unacceptable, especially where smallholder agriculture predominates, as it does in much of Asia and Africa. Believing that in this sector great disparities of income do not exist anyway, he argues that the differences that do exist can only serve to stimulate entrepreneurial activity. In a free market economy, the incentive of wealth is necessary to draw out the talents and abilities of the population. He concludes that 'Income redistribution within smallholder agriculture is therefore unnecessary, undesirable, and unachievable' (Clayton 1983: 19).

Various counter-arguments can, however, be made to show that inequality is not only socially unjust but also economically undesirable. For a start Third World élites do not plough surplus wealth back into businesses at the same rate as Europeans did during the early stages of their industrialisation. This potential capital is often spent on luxuries and imported goods. These facts should not be viewed in any unctuous way by the industrialised world, because our European forebears did not experience, as do the LDCs, the temptation of consumer goods produced by more advanced societies, since Europe was leading the industrialisation process. Secondly, gross inequalities will affect the supply of effective labour and demand for goods produced locally. If we take the figure suggested earlier as the proportion of the world's population that is only just surviving, one-fifth of a country's population will be so poor that they will have no purchasing power and, therefore, make no contribution towards a market for locally produced goods. Clearly this offers no stimulation to production, because the poor have no money for home-produced goods and the rich prefer imports. Finally, a lack of material welfare is not likely to encourage participation in the development process, only an apathetic acceptance of what life has to offer.

One explanation for inequitable growth which received wide coverage in the late 1970s, is Lipton's thesis of urban bias (Lipton 1977). Lipton argued that the persistence of rural poverty was the result of the concentration of political and economic resources in urban areas. Although subject to considerable general debate, the thesis has not been applied much in Latin America, where the considerable uneven development that does exist has been explained mostly in terms of dependency and internal colonialism.

POPULATION, UNEMPLOYMENT AND EDUCATION

Although, as already pointed out, poverty and hunger are due more to maldistribution than inadequacy in an absolute sense, the solution is often seen in terms of reducing population levels. If there were fewer poor, then feeding them would be an easier task. The population debate that has arisen over this focuses on the relationship between population and the problems of underdevelopment.

The populations of Third World countries are increasing at rates which are unprecedented in history. Every year almost 90 million people are being added to a world population of just over 5.4 billion. Eighty-one million of these additional people will be born in the Third World

(Todaro 1992: 179). In 1990 the industrialised countries represented only 23 per cent of the world's population (Third World Guide 1992: 31). In low- and middle-income countries the average rate of population growth for the years 1980–90 was 2 per cent, whereas Europe for example has a growth rate of 0.1 per cent (World Bank World Development Report 1992: 269). The explanation for the high rate in LDCs lies in the reduced death rates, and in increasingly high birth rates. The improved medical knowledge and better health facilities that have resulted from contact with the economically developed world mean that people can expect to live longer than in the past but, at the same time, there has been no reduction in the rate at which children are born. This becomes particularly apparent when looking at the age structure of LDCs. There, almost one half of the total population are children under the age of fifteen, while the corresponding age group make up one quarter of the population in industrialised societies.

But to what extent is rapid population growth a serious development problem? It is true that high birth rates are usually found in poor countries, but many economists would see this as a result of poverty rather than a cause. Historically, falling birth rates have tended to follow a rise in the standard of living of a poor society.

The main argument supporting population increase as a problem view, is that fewer people would relieve the pressure on resources and improve the quality of life for the population in general. However, a closer look at statistics for populations and resources suggests that there is no clear correlation between population density and standard of living. Bangladesh and Holland are both among the most densely populated countries, they even have similar geography with easily flooded lowlands on the deltas of the Rivers Ganges and Rhine respectively, yet while Bangladesh is one of the poorest countries in the world, Holland is one of the richest. At the other extreme of the population density scale, Canada and Australia with low population densities have high income levels, while several African countries with similar population densities suffer from famine (Third World Guide 1992: 31).

Many, in fact, argue that rapid population increase is not a major problem and that to emphasise this takes attention away from the crucial issues. Susan George takes the argument a step further by asking why it is that the rich are so keen to promote population control and to see this as a way of solving development problems. She suggests that this is a deliberate attempt to encourage adaptation to existing structures rather than alleviating poverty and inequality through changing institutions and the organisation of society (George 1976).

In economic terms, it may be argued, too, that for some countries and regions, population growth is necessary for development, mainly because people provide a labour force which is an essential ingredient for economic growth.

What is often forgotten in this debate, is the view of the poor themselves, who are frequently seen as unable or unwilling to exercise any control over the situation but who, in reality, are positively in favour of large families. A few extra hands can make a great difference to the output of a family unit, so the peasant family, who have children to help them can produce more for both the market and home consumption. Moreover, in countries where the state offers no provision or help for the old, parents must turn to their children for security when they are unable to support themselves. The greater the number of adult children, the greater the security in old age.

The main conclusion to emerge from this is that rapid population increase is not a major development problem, so that attention should be focused elsewhere, but that it is clearly linked to problem areas and therefore cannot be ignored.

Unemployment is always a very real problem in any society, but in underdeveloped countries its magnitude is emphasised by the lack of welfare systems. Where there is no social security to fall back on, unemployment causes very substantial hardship. In economic terms, this represents a distinct under-utilisation of a factor of production. Labour power is something that LDCs have an abundance yet, if unemployment levels are high, it is clearly not being used to its full capacity.

Although unemployment is considerable, it bears no comparison with underemployment which is very widespread. Underemployment refers to work that is not full-time and does not provide the workers with any real security or substantial income. The workers are not unemployed because they have some kind of job but not sufficient to support them fully. LDCs are full of people in this category, who may be street sellers, shoe-shine boys or, perhaps, the sons and daughters of farmers, whose plot of land is too small to support a peasant family. For obvious reasons, statistics for underemployment are notoriously inaccurate, but all the evidence suggests they are high throughout the region. In Peru in 1992 over 90 per cent of the working population have been classified as unemployed or underemployed (Poole and Renique 1992: 23). Unemployment is clearly linked with other problems mentioned for, with no work, poverty follows. There is a close relationship between high levels of unemployment, widespread poverty and unequal distribution (Todaro 1992: 222).

One important question that emerges is, why are there so many people without proper work, even in LDCs which have experienced economic growth? The 1950s and 1960s saw industrial development and the accumulation of capital in many LDCs and yet the modern sector failed to generate significant opportunities for labour. So, where economic growth has taken place, it has not automatically reduced poverty, nor has it increased employment significantly. The following chapter will offer explanations for this.

Education is important, not only because it leads to wealth, power and status in society, but also because it enhances the quality of life. Low levels of education and literacy are another characteristic of LDCs, with illiteracy rates as high as 74.3 per cent in some African countries, 71 per cent in some Arab states and even 90.8 per cent in parts of Asia. Guatemala has the highest level in Latin America, with 45 per cent. (This excludes Haiti whose illiteracy rate is 48 per cent.) The highest levels of more developed countries do not exceed 9.9 per cent (Spain), while the illiteracy level of the USA was 4 per cent in 1990 (The Economist *Book of Vital World Statistics* 1990: 210).

Literacy is a basic skill which makes it possible to continue to higher levels of education. This raises the question of the balance between different levels of education in poor countries. Is it more desirable to give large numbers the benefit of primary education, or to spend a substantial proportion of the education budget on higher education for the benefit of a few? The former offers limited development for the masses while the latter creates an intellectual élite and the opportunity for an educated leadership. Most Third World countries when faced with this issue have made primary education the first priority (Goldthorpe 1975).

MEASURING DEVELOPMENT

This chapter has suggested that the various problems associated with underdevelopment are clearly interlinked, but it is difficult to identify distinct causal relationships. If improvements are to be made in these areas, we need to have some way of measuring progress, some indicators which can be used to show whether or not life is improving for the poor. Although, as we have seen, there is some debate over what are the major problems, the attempt to measure development is even harder. The difficulties are very similar to those encountered in attempts at defining development. There are differences of opinion over which are the important criteria; moreover, many of the criteria are qualitative ones. How can one measure standard of living, health levels, participation in society and the quality of life in general? The only way to approach this is to measure them directly, but using indicators which can be quantified. Health, for instance, is often judged by using infant mortality or life-expectancy statistics. A wide variety of indexes are used to measure living standards, since this is such a complex factor. Average national income per person is very commonly used and does provide some sort of universal yardstick, though, as we shall see, it does have its drawbacks. Many indices relate to housing facilities, such as access to piped water or electricity, others to aspects of diet or clothing. In Mexico, since many of the peasants go barefoot because they cannot afford shoes, the proportion of the population wearing shoes is often used as an indicator.

Indices referring to housing, clothing and diet vary according to country and region.
The difficulties can, therefore, be best summed up as follows:

1. No quantitative indicator can exactly measure a qualitative criterion.
2. Bearing in mind all the component parts of development, no one indicator can successfully sum these up and represent them.
3. It is difficult to bring together different sets of statistics, measuring qualitatively different factors, into a single index of development.

(Colman and Nixon 1986)

Despite these problems, as the 'concept of development only acquires substance through a process of measurement' (Colman and Nixon 1986: 8), it is important to examine the strengths and weaknesses of the various indicators that are used.

The desire to use one single index is understandable since it simplifies the process of measurement and comparison. When analysts have relied on only one index, it has usually been economic since, as we have seen, an economic element is considered essential to any general progress. One indicator in particular has been widely accepted as an estimation of a country's level of development and that is national income or gross national product (GNP) per capita. Since GNP is calculated in every nation on the basis of the same set of rules, it has the advantage of offering information that can be compared over international boundaries. It is a statistic which measures the overall level of income or production of a society and, therefore, does identify a key activity, the provision of goods and services, an increase which reflects economic growth. These advantages make it particularly attractive to economists who see economic growth as the essential motor for development. On the other hand, the disadvantages reveal serious criticisms of relying too heavily on this one index. For a start, it takes no account of non-economic factors such as health, education, quality of life, etc. Secondly, the rules for measurement were devised in the industrialised societies and therefore make it difficult to measure income and production in poor countries where produce does not necessarily go through the market system nor is recorded in any official documents. Finally, while GNP provides a general picture at national level, it gives no indication of different income levels within countries and, therefore, may be quite inadequate in revealing poverty and inequality. Thus, for example, in the Philippines 1961–65, India 1954–63 and Peru 1961–71, GNP increased but so did inequality, showing that GNP in this case is of no value in estimating changes in poverty levels or income distribution (see Fig. 1.1). Other countries, such as Nigeria in the 1970s and Brazil in the late 1960s and early 1970s, have experienced substantial economic growth and high rates of GNP but, at the same time, supported very poor and 'backward' regions.

The latter problem can be tackled by using indicators of income distribution though comparison, in this case, becomes more complex. Income distribution tables are usually constructed by dividing a country's population into ten groups, according to their income level and then indicating what proportion of national income accrues to each group. Comparing the proportion going to the wealthiest 10 per cent with that earned by the poorest groups, will reveal the gap between rich and poor. Cross-nation comparisons of a country's wealth, showing what proportion the rich receive, also indicate different levels of inequality. Table 1.1 shows that the wealthy in poor countries receive a bigger share of the national 'cake' than they do in richer countries.

Many economists and sociologists link development to an expanding industrial centre and would, therefore, use indices relating to the growth of the sector. The percentage of labour employed in urban occupations is sometimes used to show the move from an agriculturally based society to a more industrial one.

The most accurate, but also most complex, way of assessing levels of development or measuring changes is to use a group of indicators selected from a range of economic and social areas. There have been various attempts to do this, mainly by international organisations who are concerned with aid or loans to LDCs. The number and range of the indices used reflect the aims of the exercise: that is, whether the assessment is being made for the purposes of financial assistance, improved trading relationships or welfare programmes.

A good example of this is the way indicators are used to make distinctions within the Third World in order to identify the most needy countries, which have come to be called the 'Least Developed Countries'. In the late 1960s the gap between the very poor countries in the world and those which, though in the Third World, were industrialising was becoming more apparent, and international organisations felt a need to identify those areas requiring urgent assistance. The UN conference on Trade and Development (UNCTAD) in 1968 and the UN General Assembly in November 1971, picked out the countries with severe long-term constraints on development, by using three criteria:

1. Gross Domestic Product (GDP) per capita of $100 or less in 1970. (This figure has since been updated.)
2. Manufacturing making up 17 per cent or less of GDP.
3. A literacy rate of 20 per cent or less of the population aged fifteen years or more.

Originally these figures led to the inclusion of twenty-four countries in the least developed category but, by 1991, the number had risen to forty-seven with a total population of 440 thousand million in 1990 (see Table 1.3).

The indices are in this case predominantly economic, but do include

Table 1.1 Income distribution

	Percentage of income received by lowest 20%			Percentage of income received by highest 5%		
	1960	1970	MRE*	1960	1970	MRE
Industrialised countries	4.3	6.2	5.1	19.2	14.0	–
Developing countries by region						
Africa south of the Sahara	5.2	4.1	6.2	27.8	26.4	–
Middle East and North Africa	4.9	5.0	5.3	23.9	22.0	21.4
East Asia and Pacific	5.3	6.0	6.2	–	27.9	–
South Asia	4.5	7.0	–	24.6	24.3	–
Latin America and the Caribbean	3.7	3.4	2.9	30.5	29.2	23.7
Southern Europe	5.5	4.3	5.0	21.5	23.8	–

* MRE: most recent estimate. Recent statistics are not readily available as few countries now publish this material.
Source: Derived from World Tables 1980 Social Indicators.

literacy rates as a non-economic indicator of the quality of life. If other commonly used indices were added, these countries would still appear at a low level of development, suggesting that, although the original criteria are only three in number, they are crucial ones.

Table 1.2 Income shares in Latin America

Country	Income share: Lowest 40% of households (%) 1985–89	Income Share: Ratio of highest 20% to lowest 20% 1985–89	Gini Coefficient* 1975–88
Chile	×	×	0.46
Costa Rica	11.6	16.5	0.42
Argentina	×	×	×
Venezuela	13.9	10.8	×
Mexico	×	×	0.50
Colombia	12.7	13.3	0.45
Panama	×	×	0.57
Brazil	8.1	26.1	0.57
Ecuador	×	×	×
Paraguay	×	×	×
Peru	12.9	11.8	0.31
Dominican Republic	×	×	×
El Salvador	×	×	0.41
Nicaragua	×	×	×
Guatemala	14.1	10.0	×
Honduras	×	×	0.62
Bolivia	×	×	×
Haiti	×	×	×

* The Gini Coefficient is a measure that shows how close a given distribution of income is to absolute equality or inequality. As the coefficient approaches zero, the distribution of income approaches absolute equality. Conversely, as the coefficient approaches one, the distribution of income approaches absolute inequality.
Source: Derived from *Human Development Report*, 1993: 170.

Because the World Bank makes loans to poor countries it, too, needs a working definition. In this case, they are referred to as 'low-income countries'. The criterion used here is GNP per head below $610 in 1990. There are a few notable inclusions on the World Bank's latest list: Honduras, Guyana and Nigeria (World Bank, *World Development Report* 1992: 307).

A welcome addition to these types of indicators is the Human Development Index (HDI) proposed by the UN development pro-gramme (UNDP) in 1990. The report stressed that human progress can not simply be measured by economic growth, but must also consider human well-being. The main idea of the report was to 'place people at the centre, to make them once again the masters and not the slaves of development' (*Guardian*, 25 May 1990). As well as GNP, the HDI takes into account life-expectancy, literacy and the degree of popular purchasing power, which is an attempt to look at social inequality. The Human Development Reports also include income distribution and gender disparity adjusted HDI figures, although these are not available for all countries. The HDI applies to the industrialised countries as well as the developing world. The UNDP finds sixty-five low human development countries, i.e. those countries with a Human Development Index of less than 0.5. Sri Lanka, the Philippines and El Salvador fall into this category when the HDI is adjusted for gender disparity or income distribution (HDR 1992). (See Table 1.4.)

Table 1.3 The forty-seven least developed countries

Afghanistan	Malawi
Bangladesh	Maldives
Benin	Mali
Bhutan	Mauritania
Botswana	Mozambique
Burkina Faso	Myanmar
Burundi	Nepal
Cambodia	Niger
Cape Verdi	Rwanda
Central African Republic	Somalia
Chad	Samoa
Comoros	São Tomé and Príncipe
Djibouti	Sierra Leone
Equatorial Guinea	Solomon Islands
Ethiopia	Sudan
Gambia	Tanzania
Guinea	Togo
Guinea-Bisau	Tuvalu
Haiti	Uganda
Kiribati	Vanuatu
Laos People's Democratic Republic	Yemen
Liberia	Zaire
Lesotho	Zambia
Madagascar	

Source: UN World Economic Survey, 1992.

Taking the UN's forty-seven poorest countries in the world, what do they have in common? By definition their earning capacity is very low. In 1990 the average GDP per capita was $240, compared to $980 in the rest of the developing world and $12 490 in the developed market economies (*Global Outlook 2000* 1990: 10). In addition, these countries

21

all have a heavy dependence on foreign aid and resulting debt problems. For the inhabitants, life is short, for life-expectancy rates are low, infant mortality high and, as the HDI figures show, quality of life is poor.

Table 1.4 The Human Development Index

Medium human development countries	HDI	Gender disparity adjusted HDI	Income distribution adjusted HDI
Sri Lanka	0.633	0.499	–
Philippines	0.603	0.451	–
El Salvador	0.503	–	0.488
Low human development countries			
Maldives	0.497		
Guatemala	0.489		
Cape Verde	0.479		
Vietnam	0.472		
Honduras	0.472		0.419
Swaziland	0.458	0.344	
Solomon Islands	0.439		
Morocco	0.433		
Lesotho	0.431		
Zimbabwe	0.398		
Bolivia	0.398		
Myanmar	0.390	0.297	
Egypt	0.389		0.377
São Tomé and Príncipe	0.374		
Congo	0.372		
Kenya	0.369	0.241	0.344
Madagascar	0.327		
Papua New Guinea	0.318		
Zambia	0.314		0.291
Ghana	0.311		
Pakistan	0.311		0.303
Cameroon	0.310		
India	0.309		0.289
Namibia	0.289		
Côte d'Ivoire	0.286		0.246
Haiti	0.275		
Tanzania, United Republic of	0.270		
Comoros	0.269		
Zaire	0.262		
Laos People's Democratic Republic	0.246		
Nigeria	0.246		
Yemen	0.233		
Liberia	0.222		
Togo	0.218		
Uganda	0.194		

Table 1.4   (Continued)

Low human development countries	HDI	Gender disparity adjusted HDI	Income distribution adjusted HDI
Bangladesh	0.189		0.172
Cambodia	0.186		
Rwanda	0.186		
Senegal	0.182		
Ethiopia	0.172		
Nepal	0.170		0.138
Malawi	0.168		
Burundi	0.167		
Equatorial Guinea	0.164		
Central African Republic	0.159		
Mozambique	0.154		
Sudan	0.152		
Bhutan	0.150		
Angola	0.143		
Mauritania	0.140		
Benin	0.113		
Djibouti	0.104		
Guinea Bissau	0.090		
Chad	0.088		
Somalia	0.087		
Gambia	0.086		
Mali	0.082		
Niger	0.080		
Burkina Faso	0.074		
Afghanistan	0.066		
Sierra Leone	0.065		
Guinea	0.045		

Source: Derived from *Human Development Report* (UNDP), 1993.

Early indicators of development focused on economic factors, in particular those that could be quantified. More recently there has been a move towards qualitative indices and an attempt to produce a more sophisticated picture giving a broader perspective. Although this represents a welcome attempt to include social indicators at the heart of measures of development, the HDI has not so far been widely applied.

BROAD TRENDS

Considerable research has been carried out identifying, assessing and measuring the problems of LDCs. The important point is how these are to be solved, and here the answers are varied and numerous, stemming partly from the different interpretations of the problems themselves. An important contribution was made in 1980 with the Brandt Report which brought about international discussion at the highest level, though the practical results have been limited (Brandt 1980).

The Brandt Report is the result of the work of an independent Commission on International Development Issues under the chairmanship of Willy Brandt and made up of representatives of both the industrialised and the developing world. The report deals with the world's needs in the 1980s and offers some practical solutions. Its main emphasis is on global integration: the fact that the countries of the North and the South have mutual interests. The poor countries of the world need to trade with the richer countries on improved terms and to receive aid from them in order to develop their own economies, but the industrial economies will benefit from this because a wealthier South will mean more markets for Northern goods and more trading opportunities in general. The thrust of the document is to promote trade and aid between North and South. Indeed, one of the points made is that without development the problems of the poor countries of the world will become so critical as to be disruptive on an international level. Apart from the problems already discussed in this chapter, the report deals with disarmament, energy, transnational corporations, trade and international finance, making several specific recommendations.

Willy Brandt and his colleagues were concerned with the long-term development and restructuring of the world order, but they also realised that some immediate relief was necessary to solve the worst problems. In order to alleviate the scale of human suffering in the short term, and in order to ease increasing debt problems, massive resource transfers were thought to be necessary. The proposal was that the wealthier countries should increase their government aid to 0.7 per cent of GNP, preferably by 1985 (Brandt 1980).

On the surface, aid would seem to offer some solution, depending on the extent of 'generosity' of the donor nation. However, a closer look reveals that aid is more complex, involving various disadvantages. Aid, apart from that in the form of grants, involves the recipient nation in indebtedness. Borrowing is a basic way of acquiring capital and often promotes capital accumulation to the benefit of the borrower in the long term. But, many LDCs have not achieved these successes through aid and so are now not only unable to pay off their debts, but also the interest on these loans. Aid is very rarely given without strings, either economic or political, being attached to it. From the donor's viewpoint this offers some influence over the way money is spent, but for the LDCs it may well impose severe limitations. Economically, the loan package may require a change in economic policy not necessarily desired by the recipient nation. The IMF often insists on wage controls, which may cause considerable hardship, as a condition for a loan. The assistance may be directed towards certain projects or industries, not necessarily favoured by the

Third World nation, but will provide some economic advantage for nationals of the donor country. Even food aid is by no means disinterested and can be promoted to serve the interests of the developed world as well as the LDCs. Susan George cites a 1966 US report which states 'this food aid has improved the diets of forty million of the world's schoolchildren ... *and built bigger cash markets* for American farm products. ... Over the past 12 years these rising commercial sales have brought tangible returns to the American farmer and business man' (George 1976: 199).

Political strings are of two kinds. There are those which relate to the nature of the political system of the recipient nation and, secondly, those which impose conditions as to political support in an international arena. In the first case, donor nations do not give aid to countries with ideologies which they believe to be antipathetic to their own way of life. In 1985 the USA's aid to El Salvador, which was run by a dictatorship under the guise of democracy, was massive, while instead of helping neighbouring socialist Nicaragua, the USA gave aid to the country's guerrilla enemies. Political stability is often deemed an important condition, the argument being that this is necessary for economic development to take place. There is indeed a case here, for without economic stability aid may well be counterproductive. No bank or financial organisation lends money if it feels that the recipient's situation is unstable and, therefore, likely to jeopardise the project. The problem, however, is that some of the most stable regimes amongst the LDCs have been those dominated by the military, since they can enforce law and order, and those often have the worst human rights records. Some of the most brutal situations cited by Amnesty International, such as Argentina or Chile in the late 1970s, have occurred under the aegis of military governments. Political support in the international arena usually takes the form of following the lead of the donor country in UN debates.

In practice, both industrialised and developing countries often have a preference for earned income over hand-outs. Trade, however, depends on the capacity of the countries concerned to produce appropriate commodities and is often subject to tariffs, quotas and taxes.

The nature and extent of development problems, the way they are examined and the solutions tried, vary between different countries, regions and continents of the Third World. Nevertheless, some general statements can be made about development.

Using single economic indicators, some broad trends can be identified. World Bank statistics show that in the period 1960–70, in terms of GNP per capita, the world in general experienced economic growth. The rate of increase of the lowest-income underdeveloped countries was 1.6 per cent, for the middle-income countries and the industrialised capitalist countries it was 3.7 per cent, and the socialist countries grew at 4 per cent per head per year (Griffin 1981). This, too, was a period of political independence for many small nations and so might appear to be a time of political and economic progress. However, following the arguments in this chapter, large, catch-all statistics need to be questioned in terms of their component parts. Although these figures show that there was general economic growth, the economic gap between rich and poor countries was growing wider, both in absolute and relative terms. It is clear from the statistics that, although the economies of the poor countries were expanding, they were not doing so at the rate of the rich. Moreover, the low level at which the poor countries start means that a percentage increase on that baseline will bring about an absolute improvement that is much smaller than that initiated by the same percentage increase in an industrialised country with a much higher starting point. Equivalent percentage increases only serve to magnify the absolute gap between the two.

Not only has inequality increased within the world in general but also the differences between the developing countries themselves have become greater. While some of the poor countries, such as Chad, Bangladesh and Somalia, are in real terms worse off than they were, some LDCs, including a number of Latin American states, have experienced programmes of such rapid growth that their levels of industrialisation are approaching those of some First World countries, with the result that they are referred to as the Newly Industrialising Countries (NICs). For some of these, their 1970s growth rates were higher than those of the industrialised world. The continuing growth of the NICs has challenged any dualist terminology of development, suggesting that we need language that can handle a continuum of development rather than polar types.

Inequalities grew worse in the 1980s as real incomes declined in Africa and Latin America. The 1980s have been described as the 'lost decade' for development. This might seem strange since average global growth was higher during 1980–89 than in 1960–80 (3.2 per cent compared with 2.4 per cent). 'The real problem in the 1980s was that global growth was poorly distributed' (Human Development Report 1992: 38). Looking at it another way, annual per capita growth between 1 per cent and 5 per cent is usually considered to be reasonable. The proportion of the world's population living in countries with less than that was 13 per cent in 1965–80, but by 1980–89 the proportion had increased to nearly 30 per cent (HDR 1992). For Latin America, the 1980s meant a series of economic crises. By the 1990s the situation can be summed up in the words of the Human Development Report 'between 1960 and 1989, the countries with the richest 20 per cent of the world's population increased their share of global GNP from 70.2 per cent to 82.7 per cent. The countries with the poorest 20 per cent of the world's population saw their share fall from 2.3 per cent to 1.47 per cent' (HDR 1992:38).

This pattern is also reflected within individual countries, where the gap between rich and poor has been widening. Even in the countries where income on average has risen, the standard of living of the poor has fallen. Keith Griffin has provided considerable evidence to demonstrate this point in Asia (Griffin 1981). In Brazil the richest 20 per cent of the population receives twenty-six times the income of the poorest 20 per cent (see Table 1.2). The World Bank puts Brazil's gap between rich and poor as second only to Sierra Leone (*Financial Times*, 2 October 1993: 1).

Apart from this situation being morally unacceptable, it clearly has implications for stability in the world. In the 1980s Brandt and Singer and Ansari predicted future social unrest at an international level (Brandt 1980, Singer 1982) and there is indeed considerable evidence today of social conflict caused by the tensions of global poverty. At national level, tensions are coming to a peak in South Africa, with violence causing the death of up to thirty people a night in Black townships (*Financial Times* 1 August 1993). Even in the USA, supposedly the land of equality and justice, internal contradictions have been graphically illustrated by the Los Angeles riots. The Palestinian uprising as well as Muslim fundamentalist popular movements in Egypt and Algeria bear witness to a rising number of cases of conflict due to inequality, while in Peru the Sendero Luminoso (Shining Path guerrilla movement) prepare to fight to achieve eventual total victory over their oppressors, however high the cost.

If conflict and inequality are on the increase, even during periods of rising economic growth rates, what has gone wrong? The different welfare goals, which are considered to be the object of development, are interlinked, but there is no evidence to show that progress in one will automatically lead to progress in another. So often the debate on development has focused on growth versus distribution; is it preferable to make the cake larger or to distribute what there is more equitably? Explanations for this uneven development are dealt with in the next chapter, which examines the various theories that have arisen to aid an understanding of these issues.

Finally we need to look more closely at the position of Latin America within the Third World. To what extent does it share the same problems as Asia and Africa? In what ways are its problems unique?

A quick glance at some of the tables in this chapter reveals that Latin American countries barely feature at all amongst the poorest in the world. In the UN's forty-seven least developed countries there is only one Latin American nation, Haiti (see Table 1.3) and there are only two others, Honduras and Guyana, listed by the World Bank as low-income, and these are recent inclusions following the decline in the 1980s. It is significant that the World Bank's listing of low-income countries in 1981 mentioned no Latin American countries apart from Haiti and that the HDI index produced in 1993 included Guatemala and Bolivia as well as the World Bank's inclusions for 1992 (see Table 1.4). This reflects the 'lost decade' for development of the 1980s for Latin America, which has brought one or two countries in the region into lower income groups. Latin America, therefore, falls almost completely within the middle-income category. In fact, any discussion which refers to the wealthier Third World countries, apart from high-income oil exporters, usually refers to Brazil, Argentina and Mexico. The two decades of prosperity in the 1960s and 1970s resulted from very substantial economic growth, for during these years average annual economic growth reached 5.5 per cent compared to 4.2 per cent in the developed countries (Hurtado 1986). Although the growth rate fell drastically in the 1980s Latin America had made a substantial recovery by the 1990s.

In 1990, 69 per cent of Latin America's population was urban (Feres and León 1990: 150). It follows from this that fewer Latin American than other Third World populations work in the agricultural sector (see Table 1.5).

Table 1.5 Percentage of labour force in agriculture

	Percentage of labour force in agriculture	
	1965	1985-8
Sub Saharan Africa	79.1	67.6
Arab states	63.0	39.2
South Asia	72.3	59.7
Asia and Oceania (all countries)	65.7	54.1
Asia and Oceania (excluding China)	65.7	54.1
Latin America and the Caribbean	44.1	26.3

Source: Derived from *Human Development Report* (UNDP), 1991: 468.

Latin America, therefore, is richer, more urban and has fewer people employed in agriculture than any other Third World regions – all factors which have been suggested by researchers as indicators of development. This might imply that Latin America is more developed than many Asian and African societies. However, other welfare goals have not been achieved and, in some areas, in particular equality, the situation has worsened. Table 1.1 shows that, as a region, Latin America and the Caribbean have a more unequal distribution of wealth than any other Third World region. Felix has shown that Latin American development is characterised by considerably greater income concentration than is the development of other LDCs and the industrialised countries (Felix 1983). Therefore, despite overall economic growth and overall *relative* wealth, the gap between the rich and the poor is great. Latin America also excels itself when it comes to the problems of indebtedness. Thus the major development issues that occur in a discussion of Latin America are inequality, poverty and un/underemployment. This book aims to demonstrate how increasing polarisation of society has come about and the attempts that have been made by the different sectors of society to bring about development. It is clear that social changes are taking place in that continent and that these are linked to the growth of modern capitalist society. This process has meant urbanisation and industrial growth in some areas, but poverty and hunger in others. Development issues in this case are, therefore, clearly linked with industrial growth. Although the focus of this book is Latin America, two important points need to be made about how this relates to development issues in general.

Firstly, Latin America demonstrates the fallacy of the *simple* economic-growth-leads-to-development formula most clearly because it is the most industrialised area. This is not to deny the value of economic growth as one of a number of welfare goals. This point has been emphasised because, although many may believe the 'trickle-down' theory has long been buried, there is still a tendency to suggest that Latin America, because it includes some highly industrialised countries, will benefit overall from their levels of economic growth. Instead, polarisation has taken place not only within countries but also between republics of the same region. The discrepancies in development between Latin America's newly industrialising countries and the rest is reflected in the rate of growth of manufacturers from the area, as a whole, from 3 per cent in 1960 to over 18 per cent in 1978, but with three countries in the region registering an equivalent rate from 5 to 30 per cent (Parkinson 1984).

Secondly, if low-income countries are to follow in the path of middle-income countries, as suggested by many development models, they can learn from the mistakes of others. 'Latin America must not be seen in isolation but as a microcosm of the developing world in all its diversities' (Parkinson 1984: 133).

The question, then, is how to achieve industrial growth, without increasing inequality. Can development be seen in terms that are not so centred on unequal economic growth, for example, basic needs, social equality or ecological balance? In which case, what kind of economic development is necessary to enable these goals to be achieved?

FURTHER READING

Abel, C., Lewis, N. (eds) (1993) *Welfare, poverty and development in Latin America*. Macmillan, Basingstoke.
Brandt, W. (1980) *North–South: a programme for survival*. Report of the Independent Commission on International Development. Pan Books, London.

Colman, D., Nixon F. (1986) (2nd edn) *Economics of change in less developed countries.* Allan Publishers, Oxford.
Foster Carter, A. (1985) *The sociology of development.* Causeway Press, Ormskirk.
George, S. (1990) *Ill fares the land: Essays on food, hunger and power.* Penguin, London.
Gilbert, A. (1990) *Latin America.* Routledge, London.
Seligson, M. (1984) *The gap between the rich and the poor.* Westview Press, Colorado.
Todaro, M. P. (1992) (3rd edn) *Economics for a developing world.* Longman, Harlow.
World Bank World Development Reports, UN publications and UNDP Human Development Report (HDR) for recent statistics.

The best introduction to Latin America is through the publications of the Latin America Bureau. Their special brief series presents current information and recent developments for each Latin American country, while other books focus on particular problems and issues.

2

Theories of development

Theory is often treated apprehensively by students as being some sort of unintelligible construct devised by academics to score points off each other. In reality this fear is misplaced because theory is devised to organise, in a coherent form, some body of information and, in so doing, offer an explanation. Good theory should make a set of apparently dissociated events intelligible to the observer, because it attempts to explain causes and relationships.

In order to understand the numerous problems and issues of the LDCs, a vast body of theory has emerged in the post-war period. This chapter summarises the main schools of thought and the ways in which they have developed, and offers a critical appraisal, paying particular attention to recent ideas. Although in general the theory discussed explains the relationship between the Third World and the rest of the world, those points which are particularly relevant to an understanding of the development of Latin America are emphasised. Variations in theoretical frameworks applied to different regions relate to features such as the levels of urbanisation and industrialisation.

It is understandable that social scientists, in their early attempts to explain the process of development and, correspondingly, the reasons for lack of development or uneven development, should look to the experience of the so-called developed world for ideas. Thus early development theory leaned heavily on the European experience, for which there existed a considerable body of social theory. Elements of nineteenth-century evolutionary theory, with various modifications, found their way into modernisation theory, while Marx's ideas were taken up later with the development of radical theories.

THE MODERNISATION APPROACH

Since the major change taking place in the LDCs was the growth of modern industrial societies, social scientists of the 1950s assumed LDCs would follow a similar path of development to that of the industrialised world. Moreover, in the climate of the post-war period, there was a world emphasis on reconstruction and development, which was encouraged by the burgeoning international organisations. The LDCs appeared backward and those in Latin America were thought to have a traditional oligarchic system. The USA focused its hopes of social change on an emergent dynamic middle class which would act as an agent of change. This view was shared by the orthodox communist parties who, following Marx, argued that the LDCs must go through the stages of bourgeois capitalism before they would be ready for the proletarian revolution. This general political and philosophical climate gave rise to the modernisation school of thought which developed in the USA in the post-war period.

Modernisation theorists, therefore, thinking that all societies progress along a path of development from a traditional state, constructed models of development based on the processes that had taken place in the industrialised world and then applied these to situations in the LDCs (Rostow 1960; Kahl 1970). In this way, it was argued, a Third World country could be identified as being modern, traditional or at some stage on this path of development. The theorists of the modernisation school were agreed on the general processes that made up development, though there were considerable variations in the models representing these changes.

The general argument was that the economic process of the growth of industry would produce particular arrangements in society, which are appropriate for running modern industry: in particular the growth and expansion of the middle class, as had happened in Europe. Modernisation theorists felt that in LDCs too, the growth of capitalist industrialisation would bring about a modernising middle class which would in its turn bring economic development and democracy to the poorer and more peripheral parts of their countries. Modernisation, which was lacking in Latin America, could be spread from the industrialised world to the LDCs, through the transference of technologies and consumer goods and through trade. The middle classes would import these and, with them, the values of achievement and entrepreneurship, along with other so-called modern values, and spread both these and the benefits of Western technology to the rest of society. With modernisation theory there was an emphasis on values which, because of the approach's roots in structural functionalism, were seen to be interdependent with other aspects of society. Thus economic change would alter values but, more importantly, a change in values would mean a change in the economy. These ideas are well expressed in the work of Redfield who, although writing in the early 1940s, embodied some of the fundamental notions of the modernisation theory (Redfield 1941). In general, they expected a diffusion of social, political and cultural forms associated with industrial societies. One of the major aspects of this was the 'trickle-down' effect of wealth, mentioned in the previous chapter, which they thought would follow increasing affluence from economic growth.

Modernisation theory is defined in terms of economic and cultural characteristics of the USA and Europe, and it represented Latin America's lack of these characteristics in terms of obstacles to change. The reason why the LDCs were not further along the path to development was the existence of obstacles, and development policies should, therefore be orientated towards identifying and removing these obstacles. In Latin America, one of the major obstacles was the *hacienda* which was viewed as a feudal institution blocking the spread of modern capitalism from the town to the countryside. It was this thinking that influenced many of the agrarian reform programmes (see Chapter 6).

One of the many difficulties with this notion of a smooth unilinear path of change is the existence at the same time of both modern and traditional institutions in the one society. This uneven development was explained by modernists in terms of dualism.

Dualism was an attempt to combine, in one system, theory appropriate for both an advanced society and a traditional one (Brookfield 1975; Bourricaud 1967; Lambert 1967). The argument is as follows: in many countries in Africa, Asia and Latin America, two socioeconomic systems operate within one nation's borders. The modern, based on a Western lifestyle, is materialistic, rational and individualist, and rooted in a capitalist economy. A middle class, which cultivates a Western lifestyle and culture, has emerged; industry is developing and employment is based on qualifications. The traditional is pre-capitalist and characterised by fatalistic attitudes and a labour-intensive and overwhelmingly agricultural economy. Social structure is rigid, allowing for no mobility, and traditional methods of agriculture are used. With continual change, the former dynamic sector moves further away from its traditional, static counterpart. Dualism is a term that has frequently been used to explain apparent contradictions in society and to describe this coexistence of modern and traditional in LDCs of different continents.

In the 1950s and early 1960s modernisation theory was important, not only because it was the guiding light of research in the social sciences, but also because of its practical implications. In some places, it still acts as a methodological reference point for development projects. It had considerable influence on international aid, as exemplified by the stance taken in the Brandt Report in encouraging aid for industrialisation (Brandt 1980).

By the early 1960s, however, a mood of pessimism had set in among development social scientists of all schools of thought, because of the 'development crisis', as the economic problems of the period were called. Palma has since pointed out that the error made by everyone at the time was to ignore the cyclical nature of capitalism (Palma 1981).

In both economic and political terms the 1960s failed to fulfil the predictions of the modernisation theories and, therefore, critics began to cast doubt on the value of their ideas. Despite the implementation of policies based on modernisation theory, such as the initiation of agrarian reform in order to remove the obstacle of the feudal-like estate, the economic situation was disappointing since self-sustained autonomous development had failed to take place. Politically, a power vacuum had developed because the traditional oligarchy had declined, but the new industrial bourgeoisie, partly because of the stagnation of the economy, was too small and weak to become a dominant political force. The bourgeoisie was hindered in this partly because it was reluctant to enlist the support of the working class for fear that this might lead to real social mobility.

To add fuel to this crisis in capitalist development, a major event had taken place in 1959 which had considerable repercussions throughout Latin America and influenced the thinking of scholars and politicians during the 1960s. The Cuban revolution made Latin Americans aware of an alternative form of development. It revitalised Marxist thinking in the area which had, until then in general followed a Moscow line and had not been orientated towards specific situations in LDCs.

By the 1970s it was very clear that the expectations of the modernists had not been fulfilled. Although there had been some economic growth, the 'trickle-down' effect had clearly not taken place. No real diffusion of goods from urban centres to rural areas had occurred. Far from redistribution of income taking place, the opposite had happened and the gap between the wealthy and the poor had widened (Gracierena and Franco 1978). Moreover, in many Latin American countries at this time, the political vacuum had been filled by the authoritarian state in the form of military regimes which were a far cry from the liberal democracies that the modernisation theorists had expected modernisation to produce.

These economic and political events of the 1960s had a profound effect in shaping the ideas of academics and in shifting the emphasis away from modernisation towards the dependency school. The new radical theorists were quick to point out the weaknesses in the theory itself, as well as its failures to portray the empirical reality. Modernisation theory does tend to be descriptive, a weakness admitted by some modernists themselves (Germani 1972), in that much of the so-called theory refers to describing modern and traditional stages and in between. What it never really describes is the mechanism of social change; it is forced to accept economic change without explaining it sociologically. Even the explanation of developmental problems in terms of obstacles is not very useful. Everyone would probably agree that the feudal estate is an obstacle to change, but the point is, why? By just seeing it as an obstacle, we are brought no nearer to understanding the process of change, nor to what constitutes development.

But perhaps the major criticism was the failure of modernisation theory to take into account fully enough the global situation. The obstacles to change were regarded as primarily internal, the industrialised world was seen as a promoter of industrial development. It is this last point that the dependency school focused on in particular, for they felt the reverse to be the case. According to them, it was precisely involvement in world trading and economic relationships with the USA and Europe that had brought underdevelopment in the LDCs. Many Latin American academics were critical of the use of Western models of development for the LDCs. Why should these countries not develop their own path, rather than just following in the footsteps of the USA and Europe? The North American model of development is not necessarily the 'best' for the LDCs (see Frank 1969 for useful criticisms of the modernisation theory).

DEPENDENCY

The initial response to the failings of modernisation theory took the form of structuralism. In contrast to modernisation, structuralism emphasised the differences between underdeveloped and developed countries. Structuralists argue that the multitude of links between the centre (developed countries) and the periphery (underdeveloped countries) reproduce structural differences between them by leading to further development of the centre at the expense of the periphery. In the case of Latin America, a major structuralist contribution to the social sciences was made by ECLA (Economic Commission to Latin America). As well as forging ahead in terms of theoretical development, they used these significant advances to make policy proposals. The ECLA analysts, like the modernisation theorists, stressed the importance of industrialisation for development in the LDCs, but they argued that the existing pattern of the world economy was not beneficial to the non-industrialised countries. This was because most LDCs relied on exporting natural resources to the USA and Europe and then importing manufactured goods from those countries. Since the economy of the LDCs was based on agricultural produce and extractive industries for export, and they could import the necessary consumer goods, this inhibited the development of manufacturing industry in the LDC itself. What was important for the LDCs was to achieve autonomous self-sustained economic growth, but this was being hindered by the existing pattern of trade which benefited the industrialised world more than the LDCs, since the former could develop their own industries on the bases of cheap imported resources. (This argument will be developed later in the chapter.) As the industrial world benefited, these countries were likely to impose external obstacles on countries attempting an independent path to industrialisation. The ECLA, therefore, not only suggested that the state should intervene in the economy, but also proposed specific economic policies to put its theories into practice. These policies of protectionism, exchange controls, attracting foreign investment to home industries, encouraging national investment and raising wages in order to provide effective demand, were all aimed at stimulating the production of goods locally to replace imports, that is import substitution or IS. As well as attaching importance to external obstacles to development, ECLA writers advocated the removal of internal obstacles as being necessary to promote industrialisation. One of the main obstacles for the ECLA team was the continent's pattern of land ownership.

The Cuban revolution not only instigated a profound structural transformation of society, it also severed the country's ties with the USA. Previously Cuba's economy had been almost totally based on sugar and, for this, both the market and the technology were supplied by the USA, thus tying Cuba's fortunes to the wishes of her northern neighbour. Many LDCs still have economies based on the export of raw materials and the import of manufactured goods, which makes them dependent and therefore vulnerable in various ways. For a start, natural resources in the form of agricultural produce are unreliable because forces outside society's control, such as drought, floods, blight, etc., can destroy the basis of the economy. This is in contrast to an economy based on manufacturing, where human beings are in control of the complete process. Moreover, in most LDCs only a small number of crops or yields are produced, so that a poor harvest in one will severely effect the economy as a whole. But most important, the LDCs are dependent on the advanced nations to provide the skills and technology to develop the produce and to buy the goods. If these markets do not materialise, then the LDCs will have no income with which to buy the manufactured goods which are so necessary because they are not produced at home. It might seem as if both advanced nations and LDCs are dependent in this trading relationship, but a closer look shows that the advanced nations have a much more strategic position. The exports of the LDCs, for the most part, are not essential, or they can be bought from another country. If Ecuador were to withhold her bananas, or

Ghana her cocoa, Europe or the USA could buy elsewhere or do without. However, the technology and manufactured goods provided by the advanced nations are essential to the LDCs, and leave them in a weak bargaining position. It was the difficulties of this dependent position and the way in which Cuba attempted to throw off the yoke of a dominant power that led radical thinkers to focus on global interactions – an approach which gave rise to the dependency school.

Taking its lead from theories of imperialism, dependency dominated development thinking during the late 1960s and 1970s. The vast majority of LDCs had experienced imperialist control, leaving a legacy of economic imperialism, though in the case of Latin America, political control had been cast off two centuries before.

For Marx, the main concern was for industrialised society and his interest in the nations of the periphery of the globe lay in their relationship to the central powers. It was not until the 1950s that social scientists began to use this theoretical approach with a focus on LDCs rather than the industrialised nations.

Paul Baran, referred to by Palma as the 'father' of the dependency approach (Palma 1981: 401) develops the Marxist argument further, showing how economic imperialism has an inimical effect on the economies of the LDCs, primarily because the economic surplus of these countries is being expropriated by the centre. This point has been developed by Frank in his well-known thesis that the underdevelopment of the LDCs is a result of their integration in the world capitalist system and, therefore, the solution would be a socialist revolution (Frank 1967). He argues that Latin America has been capitalist since the conquest, because its involvement in the world economy since the colonial period through the export economy has transformed it into a capitalist society itself, a point which has since been much criticised. He goes on to suggest that this integration in the world economy creates a metropolis–satellite chain stretching from the industrial nations through the towns of the LDCs and, finally, reaching the rural areas. The point about this chain of integration is that at each point the surplus generated is successively creamed off towards the centre.

It is through these metropolis–satellite links that the centre develops at the expense of the periphery, suggesting that the development of capitalism at the centre requires underdevelopment in the periphery. Frank's work in the 1960s was a major contribution to development theory in that he drew attention to the exploitative forces in operation, which generated considerable research by other social scientists. His weakness lies in the oversimplification of his views, perhaps because he was trying to cover large quantities of material across time and space.

One of the major criticisms of Frank's work is that his insistence on Latin America being capitalist since the colonial period is not well founded, a point that is well articulated by Laclau (Laclau 1971). Laclau argues that the basis of Frank's view that Latin America has always been capitalist lies in the continent's integration into a world capitalist market system. But for Marx, definitions of capitalism, feudalism, etc. were not in terms of the market, but in terms of the mode of production. It is the latter which determines the capitalist or feudal nature of the economy. In Latin America's case, there have always been non-capitalist modes of production, which have played a significant part in the economy of the region.

But Laclau points out that this does not necessarily contradict the major thrust of Frank's argument: that development of the centre has led to underdevelopment in the periphery. Laclau strongly supports this point, saying that it is not just the progress of capitalism which creates underdevelopment, but a more complex process in which capitalism interacts in a dominating way with indigenous modes of production.

Dependency theory was taken up with enthusiasm and developed by Latin American academics, who welcomed theories that were focused on the LDCs rather than taking the USA or Europe as their model of development. The dependency approach became so widely accepted in the 1970s that various schools of thought developed within its broad framework. Kay distinguishes between reformists, who wanted to change adverse terms of trade, and Marxists, whose perception of the global system as one of a super exploitation of labour led them to advocate socialist revolution as the solution (Kay 1989). Many *dependistas* pointed out that although the dependency situation was important, this analysis must be linked with one of the internal social structure. It was not sufficient to blame Latin America's problems solely on external forces, for these factors must be examined in terms of their relationships with internal factors (Dos Santos 1973; Quijano 1971; Cardoso, Faletto 1979). Dos Santos argues that dependence is a conditioning context for certain kinds of internal structure. It is not determinist, in that it defines internal structures, as Frank suggests, but it creates a general condition in which a society operates. It therefore does alter the internal functioning and articulation of the elements of the dependent formation, and creates different types of relations of dependency in different societies. These relationships may be colonial, industrial–financial or, more recently, industrial–technological, and will generate different internal structures. While Frank emphasised similarities in the effects of dependence throughout Latin America, Dos Santos stresses discontinuities and differences (Palma 1981).

In dismissing the mechanical, determinist approach, Dos Santos allows for a more dynamic role for Latin American classes which makes possible the form that dependency takes. He makes the important point that economic imperialism is only possible where it finds support among those local groups who profit from it. It can only occur where there is a coincidence between dominant local and foreign interests. Local élites compromise and dependency is structurally linked to the internal situation (Dos Santos 1973). This compromise has occurred where Latin Americans have worked as managers for foreign-run plantations or businesses, or where they have themselves invested in foreign enterprises. Clearly, their fortunes were then linked to overseas interests.

Dos Santos is primarily concerned with the dominant social class and their relationship with external forces but, as Frank has suggested with the metropolis–satellite model, the influence of dependence stretches to the poorest and most peripheral groups of society. This theme has been developed to apply dependency theory to rural areas through the concept of internal colonialism (Stavenhagen 1970). Internal colonists take, as their starting point, the position of the dualists for, while accepting that on the surface many LDCs appear to have two distinct systems, they argue that this reflects gross differences in society. These differences are not due to the social difference between modern and traditional, but precisely the opposite, that is the integration of these sectors of society in a manner in which the rural traditional sector is exploited. Backwardness is, therefore, not due to lagging development, or being left behind by the modern sector, but exists because of the relationship with the modern sector. Urban areas advance at the expense of the rural, a process which replicates the underdevelopment of the LDCs that has taken place as a result of the development of the advanced nations. The term internal colonialism implies that colonial relationships, similar to those between nations, exist within individual countries (Stavenhagen 1970).

As we have seen, one of the premises of the dependency theory is the international division of labour, in which LDCs supply raw materials for the industries of Europe and the USA, who produce consumer goods that are then imported by the LDCs. In recent years this situation has altered, which has led to some new lines of thought among the dependency school.

The classical dependency situation in the form of the enclave economy is one where natural resources are developed by foreign technology for export, thus providing no benefit to the LDC. Profits go abroad and, because of the use of technology, there is little influence on the employment sector, therefore not even providing the benefit of jobs. The effects are harmful in that the economic surplus of the nation is being lost, without the nation developing its own autonomous and self-sustaining economic growth.

This situation has changed due to the greater involvement of foreign capital with the LDC itself. Essentially, there are three main processes in what were called the new forms of dependency (Cardoso 1972). Firstly, there is a shift in investment from traditional sectors to industrial areas, in particular in technology and organisationally advanced sectors. This is only a shift, not a complete transfer, but foreign investment in primary resources has been on the decline, while it has been increasing in dynamic and expanding industries. Secondly, there has been a move towards multinational companies, which have been replacing North American businesses. Since multinational companies involve three different types of capital – local state capital, private national capital and monopoly international capital, which is under foreign control – the LDC, both on a private and a state level, becomes involved with foreign capital in a joint venture. Finally, there is a general process of increased local participation. Managers of branches of multinationals are selected from the host society, local industries become more closely linked through the provision of supplies, the purchase of finished projects and the use of technology, and local markets are cultivated more for the end products of the multinationals.

The result of this is that the most modern and dynamic sector of the LDC's economy becomes integrated with the international capitalist system, thus benefiting all involved. But this is only a small minority of society, the remainder, the backward social and economic sectors, become subordinated to the advanced one.

In summary, then, processes of differentiation and integration are taking place. Differentiation is taking place in the pattern of investment since it is going into a wider variety of both foreign and indigenous industries, in the forms of capital and in foreign ownership since there is less North American investment than previously and more European and Japanese. On the other hand, integration is taking place through the increasing identification of some sectors of the LDCs with foreign interests. This leads to the increasing polarisation of society.

The explanation of the dependency school for problems and uneven development is clearly very different from that of the modernisation theorists, and expressed in the use of the term underdevelopment. For the dependency theorists, underdevelopment is the direct result of involvement with the world capitalist system; trade with the advanced nations does not bring autonomous industrialisation and development, as the modernisation theorists suggested, but instead it hinders the process. This is for several reasons. Firstly, as the investment necessary for economic growth has an external source, the surplus from economic activity goes abroad, thus depriving the local community of an opportunity to build up its own savings and capital. Imported technology limits employment possibilities, so inhibiting the well-being of a large sector of society; the existence of multinationals in agribusiness is having the same effect on rural social structure and depriving the peasant of a livelihood, often by taking over his land. This leads to rural urban migration on a massive scale which only serves to exacerbate the urban problems of unemployment. The existence of foreign enterprises hinders the growth of a local bourgeoisie, because it makes it difficult for local industries to accumulate capital. Entrepreneurial activity is stifled since local businessmen cannot compete with the capital, the technology and the economies of scale of the multinationals. Local industrialisation and the middle class associated with this, who modernisation theorists predicted would be the primary agent of change, do not get a chance to develop if they cannot expand investment. Finally, the general process described leads to a polarisation of society, in which a small minority linked to foreign interests prosper at the expense of a vast majority who are unable to find work, while profits flow abroad. (For an excellent dependency analysis of the historical development of Latin America see Stein and Stein 1970.)

Modernisation theorists explained these contrasts between rich and poor in terms of dualism and lagging development, the backward sectors lagging behind the advanced. The dependency school explains uneven development in terms of the subordination of the backward sectors to the advance of the dominant ones. The latter can develop by accumulating capital, which means extracting the surplus of the dependent countries. The 'trickle-down' effect from the latter's wealth has not taken place because of the combination of dependence and class structure.

CRITICISMS OF DEPENDENCY

Dependency has been criticised from different perspectives. Some economists, who have inherited many of the principles of the modernisation school, disagree with it because they argue that LDCs have benefited from contact with the industrialised world.

> Far from the West having caused the poverty of the Third World, contact with the West has been the principle agent of material progress there. The materially more advanced societies and the regions of the Third World are those with which the West established the most numerous, diversified and extensive contacts. . . .The level of material achievement usually diminishes as one moves away from the foci of Western impact. The poorest and most backward people have few or no external contacts; witness the aborigines, pygmies and desert peoples.
>
> (Bauer 1981: 70)

Equality as a goal has been attacked as being discouraging to initiative and dynamism. Skewed income distribution in a free society is seen as a method of rewarding merit and productivity and thus the mechanism for drawing out the talents and abilities that exist and harnessing them to the needs of society. Inequality is therefore necessary if progress is to be made (Bauer 1981; Clayton 1983).

Many contemporary criticisms, however, move on from a dependency position, rather than totally discarding these ideas. The basic point that dependency makes about industrialised countries developing on the

41

basis of capital accumulation from the surplus derived from the LDCs provides a starting point for understanding the process of development. Nevertheless, these critics point to crucial weaknesses which flaw the logic of the argument. There is tautology built into the thinking from the very beginning. Development is defined as self-sustaining or autonomous industrial growth and yet development requires underdevelopment, as countries need to extract surplus from others in order to develop. 'Dependent countries are those which lack the capacity for autonomous growth and they lack this because their structures are dependent ones' (O'Brien 1975: 24). This circular argument leads to research which provides the historical evidence to prove the obvious (see O'Brien 1975; Booth 1985 for useful discussion of this argument).

Another problem is that the connection between dependency and socioeconomic problems of the periphery has never been adequately demonstrated. Attention has been focused on the dominant classes and their relationship with international capital, the inference being that the poor are exploited, but the mechanics of this situation are rarely identified. There is some evidence to show that foreign investment encourages capital intensive industrialisation, which leads to high levels of unemployment/underemployment. Industrialisation attracts labour from the countryside, despite the lack of real opportunities, thus aggravating the pressure on urban facilities. Agribusiness, too, encourages labour mobility, usually in the form of semi-proletarianising the peasantry. But little attempt has been made at in-depth analysis of the relationship between social class, apart from the bourgeoisie, and dependence. A notable exception here is Hamilton, whose analysis includes the working class, not just as the passive receivers of history but as an active force in the proceedings (Hamilton 1982). Are these problems solely the result of a dependent situation or are they the outcome of the growth of industrialisation and its resulting concentration of capital and machinery?

Following on this point, Lall argues that for dependency to be a theory of underdevelopment it should specify particular features of dependent economies which are not characteristic of non-dependent economies and that these features should be shown to affect adversely the development of the dependent economies. But no such factors can be clearly identified; most dependency theorists refer to features which Lall sees as being aspects of modern capitalism in the Third World. She goes on to demonstrate that it is impossible to show a causal relationship between underdevelopment and the characteristics associated with dependent economies (Lall 1975), a finding which is supported by empirical data.

Booth argues that the explanation for these limitations in theory lies in the determinist nature of much of its thinking. 'Behind the distinctive preoccupations, blind spots and contradictions of the new development sociology there lies a metatheoretical commitment to demonstrating that the structures and the processes that we find in the less developed world are not only explicable but necessary under capitalism' (Booth 1985: 776).

Warren developed one of the most comprehensive critiques, based on the view that dependency is no longer a relevant concept because uneven development in the LDCs is not attributable to imperialism, but to capitalism, which has always had a contradictory and exploitative element (Warren 1973). Warren does not deny the role of imperialism in the past, but suggests that this is now declining and no longer responsible for the problems of the LDCs. He agrees with the dependency theorists that the extension of capital into the Third World created an international system of exploitation called imperialism, but makes the point that this process also created the conditions for the destruction of imperialism by sowing the seeds of capitalist social relations and productive forces. Imperialism was, in fact, necessary to initiate the process of industrialisation and, having done so, is now retreating, leaving capitalist societies. Capitalism's contradictory and exploitative element is responsible for the problems of the LDCs. This thesis relies on a 'pure' Marxist analysis, for uneven development is attributed, not to Third World imperialist relationships, but almost entirely to the internal contradictions of Third World capitalism itself. Thus imperialism is seen as progressive because it moves history closer to socialism.

Warren demonstrates this by referring to the characteristics of Latin American economies, which were identified earlier as the new forms of dependency, only for him these represent the growth of independent industrialisation. Referring to Sutcliffe, he sets out the main criteria for economic development. 'These are that development should be based on the home market; that development should encompass a wide range of industries; development should not be reliant on foreign finance except where the underdeveloped country can control the use of foreign funds; and that analogously to having a diversified industrial structure there should be "independent technological progress"'(Warren 1973: 17).

Warren points out that the imperialist theory of the international division of labour is breaking down because of the LDCs move away from natural resources to manufacturing. In global terms, this means that the distribution of power in the capitalist world is becoming less uneven.

This view represents a departure from other criticisms because Warren lays the blame for problems squarely on the shoulders of capitalism itself, rather than dependent capitalism. Most other critics attach some importance still to an unequal relationship between developed and underdeveloped.

In addition to criticisms of the theory, economic developments began to challenge the dependency position. Even while the many dependency theorists were writing, events were taking place that would lead to a rethinking of these ideas. In the late 1960s and early 1970s the terms of trade were changing to the advantage of some of the LDCs exporting minerals and agricultural products, a situation which some were able to take advantage of and use to promote economic growth. As well as this, as their middle classes grew, some Latin American countries were starting to develop as important markets for the industrial nations. Large corporations producing cars and electrical items were finding that profits could be made by selling these within the Latin American countries. The 1973 oil crisis was particularly critical for the LDCs since it demonstrated a dramatic reversal in the dependency relationship on the part of a small number of countries. Although this situation only occurred in the case of one natural resource, it did show that dependency could be shaken and that some LDCs were capable of turning the tables on the industrial world and making them dependent, with all the vulnerability that implies, on the LDCs for a short while. The discovery of oil in countries like Ecuador, which had hitherto depended on exports of bananas, did give a boost to the economy.

These events tended to confuse the straightforward dichotomous picture of one set of countries, the LDCs, being dependent on another. Some LDCs, such as Brazil, appeared to have economies more like those of the industrialised world than those of LDCs such as Nepal, Upper Volta and Bolivia, for example. These changes led to a rethinking of the dependency approach.

Cardoso sees these economic changes as being distinctly beneficial to Latin America and, for these processes, he uses the term associated-dependent-development specifically to imply the two notions of dependence and development which, he says, have hitherto been seen as contradictory. The key institution in linking development with dependency is the multinational, since it brings into play a new international division of labour and is responsible for the injection of capital into Latin American economies, which provides the dynamics for development. Therefore, there is an increasing identity of interests between foreign corporations and Latin American states. Because of the multinational, the relationship between the developed country and the underdeveloped one no longer exploits and perpetuates stagnation but promotes growth in crucial and dynamic sectors of the economy. Development in the contemporary capitalist world requires the technology, finance and organisation that only multinational firms can supply. Cardoso, basing his analysis on Brazil, sees the growth of the multinational in Latin America as being a stimulant to development, but he also accepts that as long as the technology necessary for production comes from outside the LDC, there is still a relationship of economic dependence (Cardoso 1973).

Muñoz suggests that LDCs may be able to use these improved conditions to give them greater bargaining strength in world trade (Muñoz 1981). He maintains:

Strategic dependency does not denote simply a state to state dependency; rather it characterises a situation in which the *international* segment (that sector associated with multinational corporations) of the dominant structure of the advanced capitalist societies depends for its continued prosperity upon access to the cheap natural resources, labour and markets of, principally, underdeveloped societies.

(Muñoz 1981: 4)

The increasing importance of some LDCs for multinationals and other large corporations, which gives them greater bargaining power, has been encouraged by contemporary world economic problems. The oil crisis, inflation, trade rivalry and the international monetary crisis all sharpen the industrial nations' need for cheap and reliable sources of raw materials and labour, and expanding markets.

Muñoz is very aware of limitations that exist on the negotiating power of the LDCs, for he points out that while in theory this strategic dependency may exist, for political reasons (through greater political experience and political power) the core nations of the world economy usually have the upper hand in negotiations. The importance of his argument is that he demonstrates the fallibility of a simple dependency model and suggests that countries like Brazil, Mexico and Venezuela have some economic importance to the countries at the centre of the world economy and should, therefore, be considered semi-peripheral rather than on the periphery or dependent in terms of the world economy.

For a period in the 1970s, the terms of trade seemed to be swinging in favour of Latin America, but this did not last. One of the major problems is that although Latin America has experienced fitful periods of economic growth and development, the region has not been able to produce sustained economic development. These periods of growth have often been financed by foreign capital, thus preventing the region from generating its own investment capital, and have frequently ended in stagnation or crisis. The most recent example of this is the debt crisis, which has displayed the vulnerability for the LDCs of this type of development. Economic growth was achieved on the basis of borrowing, a very acceptable principle in a capitalist system, but the debt became so exorbitant that some countries were unable to meet the debt service payments, let alone repay the original loan. Resources must now be used to finance repayments rather than fuel new development projects.

POST-DEPENDENCY THEORY

The result of these changes in both ideas and socioeconomic development is that amongst academics, no single global theory enjoys the eminence that dependency did in its heyday. Instead, attention has focused more on specific rather than general issues, and a number of theories and approaches have arisen to explain particular problems. During the 1970s the growth of military regimes in Latin America led to a preoccupation with theories of the state, which will be examined in Chapter 9. Issues of contemporary significance, such as ecology, gender and race, have given rise to their own bodies of theory; these will be discussed in Chapters 3 and 5 respectively.

Social class

One area which the dependency school has never really got to grips with is that of internal class relations, which are so vital to an understanding of any social formation. Kay's discussion of the theories of development and underdevelopment shows how at different stages of its exposition dependency fails to make an adequate class analysis (Kay 1989). In order to remedy this and attempt to analyse the nature of class, social scientists refer back to Marx's original formula. For Marx, class relations stemmed from the mode of production, and this therefore should provide the focus for further study. Here, they are following the line of thought developed by Laclau in his criticism of Frank, and shifting the emphasis away from market exchange and the relationship between dependent and advanced societies to modes of production as the source of class relations.

The modes-of-production debate picks up on Laclaus's point about the articulation of the capitalist mode of production with pre-capitalist modes, and develops this further (Long 1977; Foster Carter 1978). The argument is that capitalism has allowed pre-capitalist modes, such as peasant production, to remain in Latin America because the former benefits through increasing capital accumulation. For example, in parts of Central America and Andean countries, peasants manage farms that are too small to support a family and, therefore, the farmer and members of his family are compelled to find seasonal wage-work on large estates to boost their earnings to a subsistence level. This system suits the capitalist estate owner who, since the peasants are desperate for work, can pay low wages and piece-rates, rather than having to maintain an agricultural labour force throughout the year during periods of little activity. The existence of infra-subsistence peasant holdings, therefore, makes possible a source of cheap labour, thus allowing greater profits for the capitalist and so contributing to capital accumulation. Articulation between capitalist and precapitalist modes of production benefits the former and leads to the dominance of capitalism over other modes but not, of course, to their exclusion. As can be seen in Chapters 6 and 7, this approach has been used quite successfully in the analyses of rural society and the urban poor.

World systems approach

The approach which has closest affinity with dependence is world systems (Wallerstein 1974, 1980; Hopkins, Wallerstein 1982). This suggests that in order to understand what is happening in the Third World we should look at the world as one capitalist system with capitalist accumulation in the core countries and surplus extracted from the periphery. The participation of other areas has been in different forms to suit the needs of capitalist accumulation. The historical development of this world system involves a stage of dependent capitalism, but analysis in terms of a dependent/non-dependent dichotomy is no longer useful. With the growth of newly industrialising countries, this theory argues that it is more instructive to use the terms core, periphery and semi-periphery, for the following reasons. Firstly, this terminology solves the problem of the dyadic relationship implied in dependency, since it has a global perspective. Secondly, it eases the difficulty of the dichotomy of developed/underdeveloped, since countries such as Brazil or Argentina may be identified as being on the semi-periphery. Finally, following on from the previous point, this form of analysis takes into account the industrialisation and development that is taking place in the Third World. This allows us to emphasise the distinctiveness of countries on the semi-periphery. As pointed out in the previous chapter, the gap between newly industrialising countries and the poor societies of the world is becoming wider. However, this approach makes use of the advances made by dependency by emphasising the importance to the development of LDCs of involvement in the world capitalist system. As with dependency, though, it is an approach which is more attractive at an abstract level than in implementation as a problematic. Difficulties arise as to which countries should be considered semi-peripheral; there is no agreement over the placement of countries in this and the periphery categories.

Postmodernism

The demise of dependency theory has meant a gap in metatheory. In its heyday, dependency offered the student of Latin America an overarching paradigm that was acceptable to most scholars and served as a basic underpinning for ideas in general. As in other academic arenas, the grand theories have been replaced by empiricism, micro-level theories and what may almost be called the antithesis of the order imposed by mega theory – postmodernism.

Postmodernism has arisen from both criticisms of large-scale theory and changes in the condition of knowledge. This set of ideas has been fed by a growing opposition to the perceived determinism of some of the grand theories such as Marxism and a reluctance to accept the general-isations of these theories. There is an increasing desire to look for differ-ences rather than similarities, to question the grand narrative and to separate out information rather than trying to order it in one major scheme. 'As a result of the computer-aided explosion in both the amount of information available to us and the variety of forms in which it arrives' (Cloke, Philo and Sadler 1991: 192) the condition of knowl-edge has shifted from a relatively modern state to an incredibly complex postmodern state. The communications revolution exposes us to a wide variety of messages from a number of different forms which can not be easily ordered in a coherent way.

This mass of information is approached through deconstruction, initially a technique for analysing texts it is used to tease out incoher-ences thus invalidating the application of general principles. The argu-ment is that in the pursuit of knowledge, broad theories conceal truths located in small-scale sites but that deconstruction can isolate them and so make them accessible. Thus spatial distributions become important, postmodernism clearly reflecting the current shift away from historicism and towards an awareness of space.

In the context of Latin America, theory has shifted from an emphasis on ideology and hegemony in the 1970s to a concern with the dis-articulation of national identities and the flexibilities of cultural and social boundaries. As architecture was the first promoter of postmodern-ism, it is not surprising that it has been addressed in some urban research. However, not all Latin American cities lend themselves to this approach. Ward offers some reasons why postmodern architecture and patterns of social organisation do not figure in Mexico City. Like many Latin American cities, the Mexican capital has not enjoyed the surplus of capital necessary to fund such developments, and during the economic crisis of the 1980s postmodern ornateness would not have been appropriate. Ward also suggests that since Mexican architects have always interpreted overseas influences in terms of local and traditional imagery, postmodern design would offer little new or significant, and there is therefore little aesthetic need (Ward 1990). Postmodernism in general has been particularly apparent in the fragmentation of the concepts of gender, race and class. 'Within each of these fields, there has been a recent move towards arguing that the central category is too internally differentiated to be utilized as a single unitary concept' (Walby 1992: 31). For instance, it has been argued that women are too divided by class or ethnicity for the category to be useful and that concepts such as 'consumption cleavages' based on housing are more useful than class when it comes to voting patterns. Recent research on Latin America applies this approach, using the new discourses, to gender and race in the search for an understanding of complexities.

Recent research into women's protest movements has criticised the existing universalising and essentialising accounts of women for not representing the interests of black women and ignoring class. This work attempts to deconstruct the homogenising accounts of Latin America that have taken place and relate gender, race and class in an analysis that reveals social complexities (see for example Radcliffe and Westwood 1993). Structuralist terms such as public and private to refer to different social and cultural domains are seen as oversimplified, thus obscuring social complexities. For instance mothers' groups in protesting about the disappeared in Latin America politicised the family in ways which shifted it from the realm of the private to that of the public, blurring the distinction between the two. Schirmer takes up the criticism of this and similar dualisms, arguing that we need to move beyond binary assumptions (Schirmer 1993). One of the problems she identifies is that the categories are exclusionary, ignoring the fact that women may have multiple strategic and pragmatic interests. A useful contribution to this debate comes from Craske (1993) who suggests a compromise position. Her notion of a public/private continuum, rather than a dichotomy, allowing for different degrees of participation between the two extremes and with many activities lying between them is valuable. She shows how Mexican women have become visible through their participation in popular mobilisations, although they tend to be the backbone of the movement rather than its head. They become more visible because they shift along the private/public continuum, as the neighbourhood, usually defined as community and domestic and therefore part of women's private sphere, is now being seen as an area of political struggle.

In an argument for deconstruction, Radcliffe focuses attention on the state. While accepting that in recent decades the state has been a very powerful force in Latin America, she warns against slipping into an account of the state which constructs it as a simple unity. She recommends 'a disaggregation of the state as a way of exploring the contestations within and around the state, and thereby suggesting that, even in bureaucratic authoritarian regimes, the "state" is contested terrain in which power blocs seek to impose an agenda of development and control in the social formation' (Radcliffe and Westwood 1993: 13). Her research shows that in Peru, images are formed and interests articulated through the dialectic of state apparatus and peasant movements. The state is not completely hegemonic: peasant groups through struggle have been able to influence state constructions.

NEO-LIBERALISM AND STRUCTURAL ADJUSTMENT POLICIES

The 1980s in Latin America were dominated by policies of neo-liberalism, which as a system of ideas has its antecedents in the classical economics of Adam Smith and the modernisation theory of the 1950s. Whereas the 1970s had experienced dependency's strong criticisms of capitalism's role towards the LDCs, the 1980s were to see a return of the modernisation view of capitalism as being beneficial. In neo-liberal thinking, capitalism is so beneficial that the purest form of capitalism is favoured, because it is considered efficient. The major aspect of capitalism is regulation through the market, so neo-liberalism focuses on the operation of a free market. The system allows the enterprising individual to thrive through a positively reinforcing cycle.

The economic aspects are underpinned by psychological arguments concerning values, aspirations and motivations similar to those supporting modernisation theory. Capitalist entrepreneurs through the market provide the dynamic for change which eventually leads to a total process of change, including social structure, political systems and culture. The success of allowing the free play of market forces can be seen in the numerous advantages of the industrialised countries, such as a high standard of living, political freedoms and democracy. In answering the question of why the LDCs are not so successful, the answers (like those of the modernists) are couched in terms of obstacles. Tradition again appears as a problem. Indigenous people are often considered to lack the necessary achievement drive. New obstacles also appear in the form of monopolies and state regulation. Monopolies which may be either industrial or those of labour such as federations of unions, limit the flexibility of the entrepreneur. State intervention in the market defies the principle of the free flow of market forces, although in Latin America the state has frequently intervened to protect, support and promote private enterprise.

The most renowned case of neo-liberalism is Chile after General Pinochet seized power in 1973. Markets were opened up to foreign competition by the removal of taxes and red tape on imports and a large number of state companies were privatised. A policy of comparative advantage followed, which concentrated on agriculture, particularly fruit and vegetables, forestry and fisheries, at the expense of industrialisation. A number of Chilean businesses suffered from competition from abroad and the shift away from urban development. Neo-liberalism was extended beyond the economic sector to areas such as education, which was exposed to a strong dose of privatisation, until 1982 when the economic crisis forced the government to adjust some of its policies (Brock and Lawlor 1985). In 1976 Argentina followed suit, and by the 1980s a number of countries, including Mexico (which had been one of the continent's great protectionist economies), were following the privatisation path.

Neo-liberal thinking has been concretised in the region in the 1980s in structural adjustment policies. This refers to the way in which national economies have had to adapt to the economic problems brought about by the debt crisis of the early 1980s and the new conditions of the world economy. Many factors contributed to the debt crisis: the oil shocks of 1974 and 1978; the consequent increase in commercial bank lending; the further economic crisis of the 1980s; rising population; the pressure on governments to provide work; and the policies of the authoritarian states. These policies have been to encourage industrial growth on the basis of foreign capital and loans, as this was considered the best way of finding the necessary capital. Between 1974 and 1978, for example, the GDP of seven of the most advanced countries (Argentina, Mexico, Brazil, Chile, Colombia, Peru and Venezuela) increased by the enormous amount of 74 per cent. But at the same time, annual net external borrowing doubled, and in the next three years, when annual GDP increased by 37 per cent, net external borrowing doubled again (Inter American Development Bank 1985). The authoritarian regimes were able, through their control of the state apparatus, to ensure cheap labour and substantial profits to attract investment, multinationals and foreign loans. Many Latin American governments supposed that the resulting econome growth would be quite sufficient to cover the debts incurred from borrowing. International trends, amongst other things, have thwarted these expectations. Interest rates have risen, due to the monetarist policies of the industrialised countries and the growing US budget deficit, which has meant that not only is it more difficult for the LDCs to meet their debt service requirements but that new lending, which would have helped them to do this, has been reduced. Moreover, the world recession has led to not only a fall in commodity prices affecting the LDCs' exports, but also protectionist policies on the part of the industrialised world, who fear competition from low-priced Third World goods. The recent GATT agreement may go some way to easing these difficulties. It has now become very difficult for many of the debtor countries to continue servicing their debts. Interest payments leapt from 17 per cent of export earnings in 1978 to 42 per cent in 1982, the year when the full effect of the external debt crisis was felt (Inter American Development Bank 1985).

The new conditions of the world economy involve what is commonly referred to as the transition from Fordism to post-Fordism, characterised by a move from a pattern of accumulation and regulation based on mass production and mass consumption to one based on the flexibilisation of labour, decentralisation and a decline in state involvement and regulation. While between 1945 and the mid 1970s Fordism took shape through the experience of the industrialised countries, in Latin America it took the form of import substitution or substitutive industrialisation, and reached a crisis point with the oil shocks and spectacular rise of international interest rates of the early 1980s. This marked the beginning of the economic recession and consequent 'lost decade' in Latin America. Globally the crises were addressed through a reorganisation of the world economy involving a move to a post-Fordist system. The main features of this are

1. technology based on microelectronics, which brings about a flexibilisation of production, thus ending the homogenisation of the labour force of Fordism and creating more specialised workers;
2. the mass market gives way to an increasingly segmented market;
3. the spread of temporary and irregular forms of employment with low levels of social protection and highly dependent on short-term economic situations (therefore a decline of the regular waged labour force);
4. extension of subcontracting and individualised wage levels; and
5. a dismantling of the welfare state.

Structural adjustment policies (SAPs) are carried out by governments in order to bring about a transition from a rapidly deteriorating economic system, in this case brought to a climax by the debt crisis, to a new system. At the heart of SAPs in Latin America lie the International Monetary Fund and the World Bank. These international funding institutions offer resources for countries suffering balance of payment problems, provided that they accept the structural adjustment terms of the funding organisation. The terms usually include

1. wage controls in an attempt to cut domestic consumption;
2. devaluation to improve the competitiveness of exports and encourage imports, implying a trade surplus to be used to pay debt interest;
3. cuts in public spending and an end to subsidies to reduce fiscal deficit and inflationary pressures; and
4. the end of government controls over prices and imports to encourage a free market to operate.

(Green 1991)

By the 1990s much of the English press was applauding the privatisation and liberalisation of the Latin American governments, emphasising improvements in economic indicators such as reduced inflation, an end to government overspending and a large trade surplus. The *Financial Times,* in May 1993, referred to Argentina after recent structural adjustment programmes, as 'the latest blue-eyed boy of the Washington international institutions' (*Financial Times* 1993: 36), with economic growth over the previous two years at nearly 9 per cent annually. Radical privatisation had been the linchpin of the government's economic programme. By 1993 this had brought in about 5.5 billion dollars, used to alleviate the debt situation and to stimulate the private sector, which the government says has since augmented investment (with some companies benefiting handsomely). Most notable of the privatisations has been the state oil company which has passed into private hands seventy-one years after it became the first government oil company in the world.

Privatisation also brought a windfall to the Mexican government's coffers. The eighteen commercial banks were sold for $12.4 billion between June 1991 and July 1992 (*The Economist*, 13 February 1993). Mexico has been hailed as one of the 1980s most enthusiastic converts to the cause of economic liberalism (*The Economist* 1993: 3). As well as the banks, large tracts of the economy have been privatised, including agriculture and the telephone company, much of it being opened up to foreign competition. After joining GATT in 1987 Mexico's average tariff on imports dropped from 45 per cent to 9 per cent. Inflation was dramatically cut from its peak of 159 per cent in 1987 to 12 per cent in 1992, and government deficit of 16 per cent of GDP in 1987 was turned into a surplus in 1992 equivalent to 1 per cent of GDP, excluding the money raised from privatisation (*The Economist* 1993).

As a result capital was flowing back into Latin America in the early 1990s. Between 1990 and 1992, portfolio flows of capital to Latin America multiplied fourfold. However, some of this may be short-term, speculative capital, which has less value for long-term development. But as Green points out, while Latin America was being acclaimed for its economic success, the first outbreak of cholera in the Western world since the 1920s had hit the region. 'Cholera is a disease of the poor which spreads through polluted wells and ditches, killing when medicines and hospitals are unavailable. Public spending cuts may win international praise, but they inflict a high price on the poor' (Green 1991: 82).

The problems of relying so heavily on market forces in a region like Latin America can be seen in the case of Ecuador. In 1992 the government of President Sixto Durán Bellén launched a stabilisation plan designed to cut the annual inflation rate (at the time running at over 50 per cent), control public spending and replenish Ecuador's international reserves, as well as initiating a package of structural reforms to reduce the size of the public sector, improve services and redirect funds into areas of social priority. In order to carry out these reforms, which involve privatisation, attracting foreign investment and dismantling considerable levels of bureaucracy, volumes of legislation had to be passed. The privatisation policies created uncertainty, exacerbating strikes and protests, which were not handled well by the politically inexperienced government. Although a development strategy with an emphasis on social measures was intended to redress the balance, there was little left in the kitty to finance this. The words of the secretary of the National Development Council, Pablo Lucio Paredes, are significant here: 'The role of the state will be first to regulate markets and ensure competition, and secondly to have social programmes – for example health and drinking water services – as well as justice and defence' (*Financial Times*, 11 December 1992). However, a major problem with such a policy is that half the population of Ecuador is outside the market, with low productivity, minimum education and no capacity to save.

The social costs of these policies have been felt throughout the region. Wage controls led to significant declines in real wages, the biggest cumulative declines in the mid 1980s being in Brazil and Mexico which had the largest debts. The resulting drop in economic activity, 'between 1981 and 1984 the Latin American product per capita declined by 9 per cent, the worst performance since 1930' (Portes 1989: 10), has had repercussions throughout society, in particular unemployment rose and the informal sector expanded. Also contributing to open unemployment is the strategy of decentralising production activities into micro-enterprises and sweatshops, which encourages the dismissal of regular employees. Standards of living were further eroded by programmes of fiscal authority, which reduced expenditure on public services. Economic adjustment processes also led to increasing income concentration.

The main burden of all this fell not on the very poorest but on the lowest paid wage-earners because of the significance attached in the adjustment process to cutting back aggregate demand.

In Peru these social costs undermined the credibility of the new president. Soon after coming to power in July 1990, Fujimori announced an economic package in response to the IMF's required programme of liberalisation of the economy, privatisation of state-held enterprises and reduction of tariff barriers.

> By February, 1991, real wages were worth only a third of what they had been when Fujimori took office. . . . A majority of Peruvians were living in absolute poverty, and only eight per cent of the adult population was fully employed. Ninety-five per cent of Peruvian workers earned less than the minimum legal wage, lacked social security and worked less than four hours a day.
>
> (Poole and Renique 1992: 151)

The negative effects on employment, wages and living conditions awaken a response in certain social groups such as trade unions, employees of state and public enterprises because privatisation leads to a decline in their numbers and professional associations as deregulation takes away a large number of their functions, transferring them to the market. It has been argued that the period of transition covered by these structural adjustment programmes leads to a temporary stage of social disintegration, which may become permanent in the form of segmentation. 'Many of the tendencies which, at times of crisis and structural adjustment, seem to point towards social fragmentation and/or disintegration finally turn out to be factors helping to shape this new social order: for example fragmentation turns into segmentation; disintegration turns into pluralism and so forth' (Tironi and Lagos 1991: 45). With the withdrawal of the state from the economy there is less emphasis on the state function of social integration. Tironi and Lagos go on to suggest that segmentation and dualism, which were seen as aspects of transition in the process of modernisation by the modernists, may become a permanent feature of post-Fordism in Latin America. The influence of Post-Fordism on the Latin American economy will be looked at in more detail in Chapter 7.

CONCLUSION

Ideas about development have changed dramatically and covered a wide spectrum in the last half century. The modernisation school saw development resulting from new internal values which, because LDCs lacked them, had to be encouraged by external factors emanating from the industrial countries. For dependency theorists, these external factors were the root of the problem of underdevelopment. Although a compromise has been suggested by Streeton, who argues that the industrialised world propagates both positive and negative impulses (Streeton 1981), in general these two schools have produced opposing sets of ideas.

The legacy of dependency that has taken up by later theorists is the emphasis on capital accumulation and the theme that the centre can only develop by capital accumulation, which is dependent on extracting the surplus from the periphery. The large profits made in Third World countries are frequently exported abroad. Would it then be possible for Latin America to extract itself from the network of international relationships in which it is involved? Kitching argues that whether or not LDCs choose a capitalist or socialist model of development industrialisation is necessary, and that it is not possible without investment capital. Since these countries are too small and too poor to produce savings from internal sources, investment must come from outside. Thus simple nationalism is, he claims, not very constructive (Kitching 1982). Problems of generalisations arise again, since a country like Brazil has pockets of great wealth and capital investment and also areas of extreme poverty.

In some ways theory appears to have come full circle, with policies of neo-liberalism picking up some of the main features of modernisation. Again, market capitalism is stressed as the source of development and values of individualism and achievement are encouraged. Obstacles such as tradition offer the main barriers to advancement. Post-modernism, however, marks a very distinct break with the process of modern rational thought and with Marxist and neo-Marxist theories such as dependency. This is a move away from the grand theory, whether pro-capitalist or not, and a move towards a very different way of thinking. To date, post-modernism has had a limited impact on research in the region, though where it has been implemented, as in studies of gender, ethnicity and nationalism, some very worthwhile work has been carried out.

The decline of a dominant global theory has stimulated a greater interest in empirical issues. Theory, however, cannot just be discarded; it shapes the choice of subject for research and is essential in providing a framework for analysis. In recent years, theoretical developments have taken different paths, relating to specific topics rather than operating at a macro-level. These ideas can be examined further by closer analysis of the empirical material.

FURTHER READING

Booth, D. (1985) Marxism and development sociology: interpreting the impasse. *World Development*, 13, No. 7.

Cardosso, F. H., Faletto, E. (1979) *Dependency and development in Latin America*. (Translated from Spanish by M. M. Urquidi.) University of California Press, Berkeley.

Frank, A. (1969) The sociology of development and underdevelopment of sociology. In Frank, A. G. *Latin America: underdevelopment or revolution. Essays on the development of underdevelopment and the immediate enemy*. Monthly Review Press, New York and London.

Kay, C. (1989) *Latin American theories of development and underdevelopment*. Routledge, London.

Muñoz, H. (1980) *From dependency to development: Strategies to overcome underdevelopment and inequality*. Westview Press, Boulder, Colorado.

Streeton, P., Jolly, R. (eds) (1981) *Recent issues in world development. A collection of survey articles*. Pergamon Press, Oxford.

Roxborough, I. (1979) *Theories of underdevelopment*. Macmillan, London.

Wuyts et al. (1992) *Development policy and public action*. Oxford University Press, Oxford.

3

'Sustainable development'

Chapter 1 addressed the question of development in terms of economic and social indicators; it is now time to look at the ecological dimension. It was pointed out that economic statistics, while providing good indicators of growth, could disguise a skewed distribution of benefits. So too these figures record the productive utilisation of resources, without revealing whether or not these resources are renewable and what lasting influence that utilisation has on the environment in general. For example, deforestation, which is usually considered as a net contributor to capital growth, not only involves a loss of resources but also has an international reputation for its ecologically damaging effects. 'From an environmental standpoint, then, GNP is a particularly inadequate guide to development since it treats sustainable and unsustainable production alike and compounds the error by including the costs of unsustainable economic activity on the credit side, while largely ignoring processes of recycling and energy conversion which do not lead to the production of goods or marketable services' (Redclift 1987: 16).

The term 'sustainable development', initially used in the early 1970s, refers to development which is environmentally beneficial or at least benign. This implies a political commitment to a programme of economic growth that will ensure a renewable resource base and a respect for quality of life in environmental terms. Thus ecologists, economists and other social scientists need to work together to ensure environmental protection that does not blunt social and economic aspirations, and to encourage economic developments within an ecological framework, which was indeed a conclusion of the Brundtland report in 1987 and an aim of the resulting Rio Summit Conference on Environment and Development. A link can be made between these two axes through the notion of livelihoods. Both development and environmental protection mean improved livelihoods for the masses (Redclift 1987).

Graham Woodgate also examines the concept of the sustainable livelihood within the context of small agricultural producers in Highland Mexico. He sees the term 'sustainable development' as a contradiction in terms and prefers the term 'sustainable livelihoods', as there is no implication concerning the possibility of continued development in the future (Woodgate 1991: 156)

Thinking about the environment within a sociological analysis has, in general, been greatly neglected. However, as this world is faced with the destruction of our planet, and the breakdown of the bio-diversity system, the urgency of such a project is finally being appreciated. The incorporation of ecological variables into the sociological analysis of the development process is the most pressing project for this reason. The tradition of this failure has its roots in the foundations of the discipline which lie in the works of Marx, Weber and Durkheim, all of whom were determined to break free of any biologically grounded social theory. Explanatory theory has therefore largely been concerned with finding non-naturalistic causes for progress in human societies (Redclift 1987: 9)

Neo-Malthusianism challenges the view that biological accounts lack credibility. This school of thought takes as its principal edict the original concept which Marx and others were involved in challenging. The essence of this principle is that population cannot exceed resources without famine or disease providing natural checks on population growth. It provides the basis of widespread conceptions, already discussed in Chapter 1, that population growth is the cause of poverty or, in this case, environmental degradation. However, from the perspective of a less-developed country the emphasis on population growth would appear to be an attempt to evade the issue of the role of the international economy in structural underdevelopment. Hardin (1968), as an extension of Malthusianism, argued in the influential 'tragedy of the Commons' that people are incapable of putting 'collective' interests before 'private' ones. In later writing Hardin and others have argued that pre-emptive, even coercive, action is required to control population and conserve resources. This reduces the role of human beings in their own development and has been described by Pepper (1984) as 'ecofascism'.

A third path, which takes nature into social analysis but which is not naturalistic, is the subject of discussion of this chapter. 'Nature', as we understand it, is in fact a social construction, and this viewpoint determines the relationship we have with nature. A rejection of unsustainable development, which was once thought of as socially acceptable, is finally taking place. What are the implications of such a rejection?

This chapter will also examine the implications for the environment, explanations for the style of development and the kinds of responses that have occurred.

UNSUSTAINABLE ECONOMIC GROWTH AND URBAN ENVIRONMENTAL PROBLEMS

If we look at the context of economic growth in the region, it would be an understatement to say that it has not in general taken place within a 'sustainable development' framework.

The 1970s in Latin America were characterised by economic growth. The average annual growth rate of the gross domestic product for these years was very high, even by developed world standards, at 5.9 per cent. This was fuelled by industrialisation programmes, invariably implying the use of modern technology. Foreign capital, usually in the form of multinationals, was generally encouraged. The most advanced technology could be harnessed, very often without the demands of environmental safeguards, so that the race for economic growth could be undertaken without any of the limitations of safety standards. Brazil offers a good example as it achieved notoriety for both its rapid rates of growth and the resulting levels of environmental destruction.

In the late 1960s Brazil embarked on a growth pole model of development. The plan involved maximising economic growth through large state investments in heavy industry (200 new state firms, many of them the largest in Brazil, were created between 1966 and 1975) and encouraging foreign capital into the dynamic manufacturing sectors. The foundations of the new economy were to be both the underdeveloped source of low-cost labour and large often prestigious, capital-intensive projects. These policies inevitably led to a dual-sector economy. The emphasis on rapid economic growth and the desire for impressive grandiose projects led to a two-tier system, with one sector forging ahead in terms of economic indices, while the other remained firmly entrenched in traditional, small-scale economic practices and in general poverty. In both urban and rural areas one sector was linked to foreign capital and geared to production for export, while the other served with increasing paucity a domestic market.

The need expressed by Congress in the late 1960s for an environmental policy was realised in 1973 with the creation of SEMA, a new agency devoted to the environment. However, SEMA has been ineffective in influencing the planning cultures of other ministries, while the development plans at national and regional levels have failed to incorporate appropriate environmental dimensions. The style of development planning is primarily dominated by macroeconomic criteria.

In the late 1970s the so-called modern sector became more and more dependent on loans and foreign capital for its growth, with the result that the changing terms of trade and rising interest rates of this period led to the enormous debt problems of the 1980s. As the burden of debt became more onerous, emphasis on exports to try to earn the much needed dollars grew, leading to an increased dependence on capital-intensive sectors which, because of their size and use of technology, were most damaging to the environment. The building of the Itaipú dam cost Brazil $20 billion, which was 20 per cent of its foreign debt at that time (Green 1991: 46).

According to Adams, in the Third World, 'pollution seems to have been part of the price paid for success in development' (Adams 1990: 115). He comments that it is precisely those cities in the Third World which are more industrially advanced that have achieved high levels of pollution. 'Cities such as São Paulo in Brazil are probably among the most heavily polluted environments in the world' (Adams 1990: 114). This figure is borne out by statistics, for example those relating to the problems of acid rain. A world resources study of concentrations of sulphur dioxide in urban areas sampled sixty-one cities throughout Africa, North America, Latin America, Europe, Oceania and Asia. During the period under study, 1973–80, São Paulo emerged as the city with the second highest average concentration (World Resources 1986). The neighbouring city of Curbatao came to be known as 'Death Valley' in the early 1980s because of the poisonous gases and effluent produced by industrialisation.

Problems also arise from other by-products of industrial processes and from the disposal of wastes. Wood-pulp plants, which are numerous throughout Latin America, are a major source of urban pollution. In Porto Alegre widespread vomiting among the population was caused by effluent, carried by the wind from a wood-pulp plant near the town. In urban areas, such as Santiago de Chile and Mexico City, a very visible dark cloud hovers over the city for much of the year. In Santiago, the snowcapped Andes that once loomed over the city can rarely be seen. The Chilean government has seriously proposed blasting a hole through the mountain chain to allow the wind through to clear way for the murk, ' "They think it's easier to move mountains than to control the bus driver" explained a resident' (Green 1991: 51). The detrimental effect on urban populations is especially felt by those whose protein and vitamin consumption levels are low.

One of the most obvious sources of pollution is rubbish, which creates a very clear spectacle of dirty streets and rivers. The great majority of Colombian municipalities for instance do not collect the rubbish, leaving two and a half million tons of waste in the open air each year. Colombia has approximately 120 million rats which, according to the World Health Organisation, contribute to twenty-six classified causes of sickness and death. The waste also reduces the levels of oxygen and destroys the eco-system of the rivers. Almost 700 tons are thrown into the Magdalena river each day which runs for 1531 kilometres through the country (*El Tiempo*, 6 June 1988).

The worst dumpers, however, are the developed countries. Many countries with tough environmental laws, such as the Netherlands, Austria, Switzerland, Germany and the USA, now ship their filth elsewhere. For example, according to Carlos Milstein, deputy director of Technology Imports, 200 tons a week of hazardous waste was being shipped into Argentina in October 1991, and local entrepreneurs are now planning to import 250 000 tons of plastics a year for incineration and dumping (Ransom 1992). In Brazil also, huge lead smelting plants recycle the lead from car batteries returned by well-meaning motorists in the North. Workers in and around the smelters now have very high levels of lead in their blood (Ransom 1992). 'It may look good locally in the rich world – but it makes no difference in global terms' (Ransom 1992: 6).

ECONOMIC GROWTH AND RURAL ENVIRONMENTAL PROBLEMS

Environmental problems proliferate as much in rural areas, being as they are in many cases directly linked to national economic policies. Environmental change in rural areas in Latin America has a history which dates back to the conquest. It has been a longstanding pattern of the South being penetrated by new agricultural production technologies and marketing which has served to push agriculture in parts of Latin America away from traditional, environmentally sustainable systems towards greater specialisation and economic dependency.

The long history of production of agricultural products for export (an early example would be the indigo and cochineal exported during Britain's industrial revolution from Guatemala to the textile factories of Manchester, or the sugar plantations of sixteenth-century Brazil) could be exemplified by the multinational fruit companies that have dominated the economies certainly of Central America for many decades. Bananas, for example, are still a booming business in Panama, Costa Rica, Honduras and Guatemala. The US fruit companies Dole, Chiquita (formally United Fruit) and Del Monte still own thousands of hectares and they started to expand in the early 1990s because of improved prospects in newly commercialised Eastern Europe and the projected removal of trade barriers in the EC. Costa Rica is one of the largest banana exporters in the world, yet very little of the profit remains in the producer country and 'as profits vanish abroad, the economic dream of easy money turns into ecological nightmare' (Lewis 1992: 289).

The current banana boom is a model of modern unsustainable development. Original sustainable patterns of agriculture relied on bio-diversity to produce maximum outputs, whereas intensive mono-agricultural operations are highly dependent on the use of pesticides. 'Costa Rica's pesticide use is equivalent to seven times the world's per capita average, and there are 250–300 cases of pesticide poisoning involving agricultural workers reported in Costa Rica each year' (Lewis 1992: 289). Many of the pesticides used are banned in the USA. The pesticides are dispersed into the local environment and massive mortality cases due to acute poisoning in wildlife have been reported, as well as among domestic animals such as cattle and bees. Up to half a million fish were found dead in the banana-producing region of Costa Rica in July 1990 (Lewis 1992).

When in 1985 United Fruit stopped production of the Pacific region, 6500 hectares of previously high quality soil were left poisoned with copper and unusable for almost any kind of agricultural production. Over 80 000 hectares of banana plantations have been abandoned along the Caribbean coast since 1979 due to the spread of Panamanian banana disease which has resulted from intensive use of chemicals. Such devastation produces a sizeable landless and redundant workforce (Lewis 1992).

The Brazilian Amazon serves for analysis of the most recent experiences of colonisation and destruction of pre-existing social systems and soil fertility in the general pattern of rural environmental exploitation in Latin America. To make more land available for national economic policies, large expanses of forest land were cleared for developmental projects (see Table 3.1). The purpose is social and economic progress, but even this is dubious and the environmental destruction caused is certainly not in doubt.

Destruction of the forest poses a serious threat to the lifestyles of the inhabitants of the forest (15 million people live in the Amazon, Cleary 1991: 129). This will be discussed later; suffice it to say here that deforestation also destroys plant and wildlife unique to the area. And, ethical and aesthetic implications aside, there are also severe economic costs. As plant and wildlife disappear, or as their genetic diversity is reduced, then potential advances in medicine and agriculture are also affected.

Some idea of the potential loss in this discussion for the future of the rain forest and all its inhabitants is given by Brian Boom of the New York Botanical Garden, who worked with the Chácobo Indians of Bolivia. He studied a sample hectare of rainforest which contained 649 trees of ninety-four species with a diameter of more than 10 centimetres. The Indians had uses for seventy-six species (78.7 per cent) either as food or in building, medicines or commerce (Prance 1990).

Much of the land is turned over to cattle ranching, which does not involve the best use of Latin America's most abundant factors of production – labour and land – as ranching has a low demand for labour

Table 3.1 Landsat surveys of forest clearing in legal Amazonia

State or territory	Area sq km	Area cleared (sq km)			
		By 1975	By 1978	By 1980	By 1988
Acre	152 589	1 165.5	2 464.5	4 626.8	19 500.0
Amapá	140 276	152.5	170.5	183.7	571.5
Amazonas	1 567 125	779.5	1 175.8	3 102.2	105 790.0
Goiás	285 793	3 507.3	10 288.5	11 458.5	33 120.0
Maranhão	257 451	2 940.8	7 334.0	10 671.1	50,670.0
Mato Grosso	881 001	10 124.3	28 355.0	53 299.3	208 000.0
Pará	1 248 042	8 654.0	22 445.3	33 913.8	120 000.0
Rondônia	243 044	1 216.5	4 184.5	7 579.3	58 000.0
Roraima	230 104	55.0	143.8	273.1	3 270.0
Total	5 005 425	28 593.3	77 171.8	125 107.8	598 921.5

Source: Derived from Cleary, in Goodman and Redclift 1991: 123. Source cited: Fearnside (1986) and World Bank estimates.

and produces a relatively low productivity per acre. Evidence for the economic rationality of this behaviour comes from the work of Christopher Uhl (1984, 1985, cited Nugent 1991) in eastern Pará. Uhl has been working in Paragominas, an area of Lower Amazonia well known due to the rapidity of forest clearance. The main reason for the high rate of forest clearance is the generosity of the Brazilian government in rewarding investors with tax incentives, not for actually producing exportable beef but for being seen to be doing so through clearing the forest. More than 23 per cent of Paragominas had been cleared by 1989. Officially, 3200 sawmills are operating in the state, many working twenty-four hours a day.

In Uhl's study of seven municipios, 99 per cent of cleared forest land was used as pasture for 1 797 000 head of cattle. The annual production of beasts represented as weight per hectare was 65 kg, of usable meat 22 kg. On the same amount of land required to sustain one steer, farinha worth seventy times as much could be produced. A Castanha tree allocated 6.25 sq. m of land, the amount required to produce one hamburger, annually produces 30 kg of nuts with a protein value rated at 21 per cent. A steer requiring more than one hectare produces only 22 kg of meat with a protein rating of 19 per cent. Therefore, when compared with the value of agriculture utilising the same area, the 'beef rationale looks feeble' (Nugent 1991: 146).

The hamburger connection has been strongest in Central America. Nearly 40 per cent of the forest cover in Central America has been destroyed. This land is now pasture for cattle which supply cheap beef to North America's fast food industry. For more than thirty years, the USA's appetite for cheap, imported beef has been a critical factor in the future of Central America's rainforests (Nations and Kormer 1987). In

63

the meantime, timber-loggers, with their negligent harvesting techniques and clumsy equipment, grossly disrupt tropical forest ecosystems. In addition to trees lost to harvesting, as much as half the
remaining forest is destroyed beyond recovery (Myers 1981).
    The tree-fellers are usually people affected by severe poverty acting to
promote individual survival. Other examples of the poor destroying the
forest due to economic necessity abound but the most famous one is that
of the *garimpeiros,* gold prospectors who invade the Amazon, most
recently in Yanamami territory, like a plague in search of valuable gold
nuggets. The destruction which they cause is unprecedented.
    David Yanamami, a leader of his people, describes the death the gold
prospectors brought in their wake:

> They are destroying the rivers with mercury [used to separate gold from its
> ore], the oil from the machines and the mud. The fish that the people eat, the
> water that they drink makes them sick.
>   The gold prospectors leave the land full of holes. Water collects in them
> and stagnates. This forms a breeding ground for mosquitos. The prospectors
> brought malaria. It's spreading all over the territory. For example, in the
> Xidea region there are many landing strips. From there they span out on foot,
> creating a network as they spread out over the hills and valleys. [The regions
> of] Parafure, Maita, Xamatare – these places are all contaminated: everyone is
> sick with malaria, with flu, pneumonia, skin diseases.
>
> (Cited in MacDonald 1991: 24)

Many developmentalists believe following on from this and other
arguments noting the cultivation of poor soil due to unfair land distribution, that inequality is to blame for many of these problems, and propose
addressing them through the extension of education and health services,
and the creation of employment (e.g. World Resources 1992–93: 30).
However, this diverts attention away from the major issue which is the
involvement of international governments, private banks and multilateral financial institutions in influencing Latin America's internal
policies on development, the environment, and tribal peoples. Inequality
is a repercussion of the pattern of development and forms part of a
spiralling vicious circle which leads to further and further expansion and
degradation. To attack poverty in such a way is like putting a bandage
on a cut, it does not search out the axe-wielder.
    Two major projects which were funded largely by international
institutions, the Grand Carajás Programme (which has already brought
many changes to Eastern Amazonia) and the Xingu Basin Hydroelectric
Programme, are recent examples which represent major sources of
environmental degradation and population displacement. They were
funded largely by international institutions. To the mining companies
the *garimpeiros* act as shock troops clearing the gold fields of incon-
veniences. Hence the very enthusiasm for the environmental issues in
the North sometimes creates intense suspicion among the poor.

One of the major obstacles to Brazil's drive for industrialisation has been the lack of an indigenous source of energy. The energy substitution programme of PROALCOHOL was aimed at growing sugar cane to produce alcohol to be used as an energy source. During 1975–85, when the heavily subsidised programme was launched and underway, the area used for sugar cane doubled and yields increased 140 per cent (Mendes 1987). The increase, however, was at the expense of domestic food production, often in areas of low nutritional levels and resulting in an increased demand for imported foodstuffs. Moreover, the waste product from the PROALCOHOL programme annually creates toxicity levels equivalent to the untreated sewage of two or three times the entire population of Brazil (Guimarães 1989).

Other attempts to solve the energy problem have harnessed hydro-electric power, which have brought, as mentioned above, environmental and social catastrophes in their wake. These hydroelectric projects reflected as much Brazil's need to show the world that it was an economic success story as its need for energy. Itaipú was the largest hydroelectric project in the world. For five years the Brazilian construction companies poured enough concrete every day to build a 350-storey building, the construction contract itself weighed 220 lbs (Green 1991: 47). But it was built at a time when Brazil had almost achieved a surplus production of electricity. The construction of dams for hydroelectric power not only destroyed plant and wildlife unique to the area (the Tucurui dam flooded 216 000 hectares in 1984, killing 2.8 million trees which were first sprayed with chemical defoliants: Green 1991: 47) but also led to the displacement of many tribal Indians. From the 243 000 hectares of land that were flooded by the Tucurui reservoir, up to 20 000 people were dispossessed (Treece 1987: 6). In order to clear the massive areas needed for dam building, whole tribes have been removed, often with the promise of resettlement, but usually resulting in totally unsatisfactory new habitations. The drive for so-called development has had in Brazil quite a profound influence on the livelihoods of a number of indigenous groups. The most widespread effect has been displacement of tribes from their native lands, though in other cases their health, livelihoods and cultures have suffered. The Parakana were moved four times to allow for the building of the Tucurui dam before they were finally resettled on a very small piece of land (Brady 1992). As a result of their movement they became disorientated, neglecting their subsistence base and making themselves dependent on government handouts. Health problems added to the difficulties, as defoliants used in preparing the area for flooding have caused sickness because unused drums of the chemical were dropped in the rivers (Brady 1992).

It is not only dam building that disrupts settlement patterns, but also extensive mining. Thousands lost their lands because of the Grand Carajás Project's programme of large-scale mineral extraction. Fifteen thousand Indians were effected by this iron ore project, including the Guajá of Maranhao, Brazil's last purely nomadic, hunter–gatherer people (Treece 1990: 277).

Cattle ranching, as already indicated, is clearly linked ultimately to North American hamburger chains, and mining is similarly linked to the contracts which supply Europe with iron ore and contain favourable pricing conditions which will contribute to preserving the competitivity of the European steel industry (Treece 1987), with the result that Brazil's resources are being exported to the benefit of the industrialised world.

TOP-DOWN 'TECHNOCENTRIC' SOLUTIONS

Some authors (Pearce, Markandya and Barbier 1989) advocate pricing of environmental losses and the development of technology which would reduce degradation. They suppose, amongst other things, that recognition of true economic value is necessary for proper natural resource management. In the view of some neo-classical economists the environment is largely thought of as a commodity. They have mainly been devoted to the refinement, expansion and implications of this view (Pearce 1985, cited in Redclift 1987). This approach has been popularly adopted by the World Bank, the United Nations and the ODA. As international interests become more entrenched so they seek new technologies to reduce risk and further their economic objectives.

Development is rarely, however, always economically rational, whether or not the environmental costs are taken into account. Brazil, for example, while destroying forest at an extremely fast rate for the production of export-orientated beef profits is not actually a net exporter of beef. In fact 40 per cent of Brazil's beef is imported (Hecht 1985).

To overcome the economic impasse of the 1990s the medicine prescribed by the World Bank is more of the same. Latin American countries, the bank argues, have no choice but to increase the mobilisation of domestic resources to further expand trade in agricultural products. Further large-scale irrigation projects funded by private capital for agro-export production are recommended. Accompanying this is some rhetoric concerning the safety of the environment. The nature and degree of change proposed by those who would rely on market forces to produce 'sustainable development', would be in the form of modest institutional changes, thus, 'they believe that the existing economic systems need only minor modifications in the form of constraints and incentive systems to force or cajole decision makers to "do the right thing" environmentally' (Miller 1991: 41). Such advocacies of environmental management have in fact largely been discredited following the fiasco of the Earth Summit at Rio. Pledges were made at the Earth Summit to throw money at the situation, but these have failed to materialise precisely because it would probably be a waste of time.

Even five weeks before the summit it was being branded a failure in national newspapers due to its narrow framework which 'crippled the process' (*Guardian*, 24 April 1992: 25). Speaking on behalf of Action for Solidarity, Equality, Environment and Development (ASCED), it was reported that the summit failed to recognise the crisis and fundamental breakdown of our society.

Within five months of the Rio Summit all the promises of financial aid for the environment had turned to dust. The four areas in which the prosperous world undertook to help the developing world are now dead in the water (*Guardian*, 16 October 1992: 23). Firstly, solutions in the area of Third World debt were covered, but prospects of write-offs have never even been discussed. Secondly, in the realm of free trade the advocates of GATT and the Uruguayan round whose free trade talks are seen as being central to the success of the summit have been attacked along the lines of 'How can more unregulated trade do anything to protect the planet?'(*Guardian* 24 April 1992: 25). The third proposal (an action programme for 'sustainable development', which pledges finan-cial support) has been strongly criticised by journalists: 'this comes as a crashing disappointment . . . it extols international trade and global economic integration as self-evidently good and calls for the combina-tion "Sustainable development" as if chanting this mantra frees us from the obligation to define it and absolves us from the sin of robbing the future' (*Guardian* 10 April 1992). On the same issue, the $3 billion earmarked for the Global Environment Facility, and promised by John Major, has not been forthcoming; nor is it even up for discussion. The fourth proposal is also a financial aid project, including $5 billion pledged by the World Bank to a sustainable development fund, but this is also unlikely to materialise according to sources within the bank (*Guardian*, 16 October 1992).

Yet governments and industry are still using the same traditional arguments. The North American Free Trade Agreement (NAFTA) was passed through Congress in November 1993. Its implementation will mean the phased abolition of tariffs on 99 per cent of goods traded between the USA and Mexico (*The Economist*, 13 November 1993). On examining the track record of many US companies now based in Mexico along the border, NAFTA is a recipe for environmental disaster. The problems are enormous. Since 1969 large numbers of manufacturing plants have set up in Mexico's border cities under the *maquiladora* programme that allows the import of components and export of finished products, free of customs duties, to the USA. Though *maquiladoras* are obliged by law to ship hazardous solid waste back to the USA, illegal dumping on the Mexican side is common. For many months of the year

67

a pollution cloud hangs over Juarez and El Paso, its Texan neighbour across the Rio Grande. Colonia Moreno, on the eastern edge of Ciudad Juarez, is now an environmental disaster. With reference to a US company, Presto Lock, a local residents' leader said: 'these people throw their waste into the streets. Each evening there is a smell like paint thinner, but worse. People complain of headaches, vomiting and skin allergies. . . . In July, a girl suffered burns to her feet while playing in the street' (*Guardian*, 19 November 1993).

Yet Mexico still argues that trade would generate more cash to clean up industry through economic growth and trinational co-operation. It is the traditional argument of industry and politicians. ' "Rational" environmental management may make the world safe for development; however, it does not make the environment safe for the poor and their livelihoods' (Redclift 1987: 172). Economic growth does not lead to trickle-down benefits, but does lead to impoverishment due to the absorption of natural resources into the money economy.

Meanwhile, NAFTA will also have far-reaching effects in other parts of the country. The pursuit of the free trade objectives will also have grave effects for farmers in Mexico. Economists in both Mexico and the USA predict that if the grain companies are successful in their efforts to force open the Mexican market, the price will fall dramatically, compelling one million or more families to leave their land (Ritchie 1992). The result – the people who own the knowledge of how best to maintain the earth for future generations will be disenfranchised from power over the pattern of the land's development.

Conservation issues have been at the forefront of recent projects of environmental management leading to the creation of such schemes as the promotion of conservation parks or biosphere reserves. Some of these are promoted by Debt for Nature swaps (see below).

An example of the conservation park solution that has been tried in Latin America is the Santa Rosa Park, encompassing 750 sq. kms, set up in Costa Rica to restore forestation and unique plant-life to the region. Key to success is the notion of a farming population, with a good knowledge of the symbiotic relationship between man and nature, taking on the 'policing' function of the park. They live on the circumference, thus in the course of maintaining their own land they prevent poachers or trippers, whose carelessness might start fires, crossing their land into the reserve.

However, other National Parks in Costa Rica (their wildlife parks cover over 11 per cent of the country's land area and have become international tourist attractions) have run into problems. Guanacaste National Park is a tropical dry forest threatened by tree-cutting and badly degraded by cattle ranching. Gold miners have been working in the Corcovado National Park, part of the Osa peninsula on the Pacific coast for thirteen years; Monteverde cloud forest is overrun with illegal loggers. Landless farmers are encroaching on many other of the country's parks and reserves (Mahony 1992: 103). Conservation parks are not a cheap solution. Most of the funding in the Santa Rosa Park case comes from philanthropic foundations and private donors, though some government financial support has also been available (*Guardian*, 22 November 1988).

John Carrier (1991) explores the political ideas which lie behind the environment programmes in Central America. He describes the Costa Rican experience as Green rhetoric from a cynical government and asks whether 'sustainable development' requires a change in the accumulation model for Central America rather than the establishment of protected areas. Despite appearances to the contrary, environmental planning is frequently toothless. Examples are not difficult to come across. Under the action programme of the European Community for the environment, most environmental planning 'has no binding effect on Member States. (They) may well adopt it somewhat cynically for this reason. Good environmental rhetoric has no political costs' (Sandbrook 1982, cited Redclift 1987: 138).

Public concern with international environmental problems has, more often that not, concentrated on issues of species survival and conservation rather than sustainability. The effect of land controlling conservation is actually to influence the place of development, *not* its occurrence. The result is often pollution havens in areas of less effective control.

Conservationist ideas have been adopted by Western advocates of Debt for Nature swaps. The idea was invented by Thomas Lovejoy who recognised that external debt was increasing pressure on the governments of developing countries to produce for a foreign market using the most economical methods, i.e. ones that often resulted in environmental degradation, in order to comply with debt repayment schemes. The effects that these agreements have on the overall debt is, however, negligible, providing only illusory relief and precipitating many other problems as well. Besides legitimising the debt at a time when the unconstitionality of the debt has been raised by authorities in Brazil, the Philippines and internationally, mechanisms which exchange foreign debt for environmental benefits are not consistent with democratic management of natural resources. 'Rather it reaffirms the creditors' political and economic domination over the debtors, within a development model which commercialises life in all its aspects' (Mahoney 1992). These are the major conclusions reached at a seminar on debt conversion as part of the Forum on Debt and Development in September 1991. The seminar was attended by environmental groups from Latin America, Europe and the USA. They went on to note that biodiversity itself is at risk from such projects as it becomes increasingly an object of interest to science and industry. The mapping of gene banks may be one of the objectives hidden behind the good intentions of the conservation and research projects allotted to debt conversion (Mahony 1992: 102).

Others promote what has been called 'bioeconomics'. This aspect of environmental management is in its very initial stages. It is still concerned with techniques rather than policies. Environmental management, biotechnological and genetic engineering, all of which can come under what O'Riordan (1981) has labelled 'technocentric' solutions, are all responses to the contradictions imposed by new technologies on the natural resource base. They are not solutions to this contradiction. They are therefore unlikely to have a positive effect and may well lead to further environmental degradation.

BOTTOM-UP SOLUTIONS

Indigenous people as subjects not objects in the process

Indigenous organisations and others (Nelson 1992) have argued that the ethnic groups affected should participate in environmental discussion and planning. One of the main causes of concern to the Indigenous people is that while the damage done to plants and trees has given rise to much international pressure, there has not been the same interest in the sufferings of people. These feelings are well expressed by COICA, the Coordinating Body for the Indigenous People's Organisations of the Amazon Basin, who say:

> We are concerned that you have left us, the indigenous peoples, out of your vision of the Amazon biosphere. The focus of concern of the environmental community has typically been the preservation of tropical forest and its plant and animal inhabitants. You have shown little interest in its human inhabitants who are also part of that biosphere.
>
> (*Guardian*, 3 August 1990)

The COICA wants to participate in the discussions about the Amazon, arguing that its members have never given up their power of representation to the international environmentalist community.

Certain intellectual positions agree with this general point:

> Development is too closely associated in our minds with what has occurred in western capitalist societies in the past, and a handful of peripheral capitalist societies today. To appreciate the limitations of development as economic growth, we need to look beyond the confines of industrialized societies in the North. We need to look at other cultures' concepts of the environment and sustainability, in historical societies like that of Pre-Columbian America, and in the technologically 'primitive' societies which present-day development serves to undermine.
>
> (Redclift 1987: 4)

The Indigenous people's land rights, integrity and superior wisdom concerning the forest have, however, been disregarded in post-colonial society. This can be understood by examining the background to ethnicity in the region – a discussion which cannot fail to engage with issues of racism, 'otherisation' and ethnicities.

The visions and myths engendered in the First World's contact with Latin America were many. The Amazon, for example, has ignited many dreams of symbolic content and mysterious allure due to its supposed immensity and the centuries of myths and silences. These fantasies have imposed upon the Amazon preconceptions that have exacted a heavy price: a refusal to permit the Amazon to tell its own story (Cockburn and Hecht 1989: 11).

According to consequent official policy, the Indigenous population was to become Brazilian via miscegenation, acculturation and integration (Winant 1992: 190). The existence of the opportunity to 'whiten' oneself was taken to mean that the 'other' would not, in the ordered world of the future, exist. Those who maintained their Indian status, like the tribal Yanamami peoples, remained outside the nation and (have recently been) subject to genocidal practices which have annexed their lands and stolen their livelihoods (Radcliffe and Westwood 1993: 8; MacDonald 1991).

Stavenhagen has described this process as internal colonialism, for many Indians lost their lands and were forced to work for ethnically distinct strangers and compelled to participate in a new monetary economy, thus becoming subordinate to new forms of political domination, since they no longer had the relative political independence of the communities (Stavenhagen 1970).

Stavenhagen's words were intended to apply to the Indians of Central America whose societies and cultures have long been influenced by the violent forces of market competition and international commodity price fluctuations. It was never imagined that the same could be applied today, to the last isolated communities of Amazonian Indians who are being taught these same lessons of international economics with a swiftness and brutality that are breathtaking (Treece 1990: 264).

The history of the experience of Latin America's Indigenous communities seems to be repeating itself, but this time with the implicit involvement of international capitalism which influences the internal policies of developing countries. During the debt crisis this process has been exacerbated and programmes and policies are limited by conditions laid down by international capitalist markets.

The 'otherisation' of the 'chaos' of the forest and its keepers (the Indigenous peoples) has been translated into or defined within economic rationalism, and the prejudices are reinforced by the conditions of the international capitalist market. For this reason any environmental programmes are underfunded, undefined and scarcely ever evaluated. Under the previous military constitution of Brazil, Indigenous people were recognised as legal minors and wards of the state Indian Agency (FUNAI), which was simply the agency responsible for executing the policy drawn up by the military government (MacDonald 1991: 27). Benefits are therefore usually intangible, while corruption due to powerful economic interests opposed to environmental measures are very real indeed.

An example of the corruption of the Indian Agency to avoid land demarcation responsibilities is their use of an internal statute which distinguishes between Indians and acculturated Indians. The latter group is said to have integrated into white society and so loses all rights to have its land demarcated (MacDonald 1991: 29). FUNAI has stipulated that if a non-Indian object is found within an Indian community, the community is in the 'process of acculturation'. Equally, this is so if it trades its surplus with a non-Indian community. Even the discovery of a matchbox would, therefore, theoretically be enough, in an extreme case, for a community to lose its claim to respect for its traditional rights, land and customs.

It is often argued that acculturation has drawn the Indigenous people from their roots and separated them from that ecological knowledge. Many studies, however, show that even where acculturation has taken place, communities still act to preserve the environment. The Kaxinawá people, for example, who had scattered from their land at the turn of the century after the death of their chief, fell prey to rubber barons in search of cheap labour. After decades as landless labourers, they organised meetings with the Brazilian government and the Indian chiefs in 1978, requesting legal titles and protection of land. It took another nine years of often violent conflict before the land was demarcated, but now this group is one of the strongest in Brazil, organising in defence of their traditional values (MacDonald 1991). Yet among the Kaxinawá community at Paroá are Indians with red or blonde hair. They wear jeans and T-shirts and travel in canoes with outboard motors. What defines them as Kaxinawá is their acceptance by the community as Kaxinawá, not their hair colour or blood type (criteria which FUNAI has used in the past) or even their economic activity. 'There is a growing recognition that there is no such thing as a pure and static culture among Indians, any more than there is among Whites. Indian cultures evolve, assimilating influences from outside without loosing their own identity' (MacDonald 1991: 28).

Similarly, the Cofan in Dureno, who have been forced into a dependence on the market economy, are now involved in the capitalist production of, amongst other things, coffee and handicrafts. Nevertheless, the community is determined to protect what little forest it has remaining against the incursions of companies and colonists. The Cofan community organises itself to clear a path around the reserve in order to deter colonists from invading (Nelson 1992). This activity could be developed into Indigenous people adopting a policing-type function, in which they could use to advantage their intimate knowledge of the environment. The suggestion is that indigenous people could provide the vital link in the conservation process.

The Indigenous people have a vested interest in conserving the forest eco-system and biological resources on which their cultures and liveli- hoods are built and sustained. Respect for the tropical rainforest is deeply embedded in Indigenous cosmology and communal ideology; local inhabitants also have a very sound knowledge and understanding of ecological balance. To achieve a sustainable situation, the knowledge of this community, who are the primary caretakers of the forest, could be utilised.

The implications of accepting this point of view, however, are greater than one might at first imagine. It does not mean simply researching Indigenous uses of the plant and wild life, or learning the farming techniques that are quickly being lost or forgotten, but rather taking on board a totally different cultural conceptualisation of the relationship which people have with the environment. It has been argued that there must be a cultural shift whereby the centrality of the environment is recognised. It would be imperative therefore that we 'see' nature differently.

The rubber-tappers have shown clearly that people can develop the forest in ways that do not destroy it as a resource for future generations. The sustainable methods they use involve working trails called *estradas*, which link about 150 rubber trees, in the forest. The trails are worked in rotation leaving each to recover before returning to bleed the trees for their raw latex. The survival of these trees depends on the survival of the ecosystems around them. Each tapper works alone, cutting his own trails, working at one with nature. They work for eight months of the year at this; the rest is spent collecting Brazil nuts.

Women as subjects not objects

Another social group who have often been promoted as the possible providers of a link to a more sustainable practice of production are women.

Advocates of ecofeminism analyse the dichotomy between women and nature on one side and men and domination on the other:

> There is an intimate association between women and the environment. It is precisely this assumption that has allowed the domination of both women and nature. . . . There is a new insight, however, provided by rural women in the Third World, and that is that women and nature are associated not in passivity but in creativity and in the maintenance of life.
>
> (Vandina Shiva, cited Rodda 1991: 5)

It is clear that 'women (do) control most of the non-money economy especially in developing nations through bearing and raising children, and providing much of the labour for household maintenance and subsistence agriculture' (Momsen 1991: 93). It is they who understand the impoverishing effects as the natural resources they utilise are drawn into the money economy.

This fact is being recognised by international development banks and agencies who now devote much study to the female role in the economy, advocating education and financial assistance for impoverished women. This can be interpreted as a call to integrate women into development and the international economy. Interpreted in this way, however, development only serves to exacerbate women's poverty and entrench their dependence on men (Simmons 1992).

García Guadilla has other reservations about the benefit of eco-feminism. She argues that the political potential of these groups is limited because environmental issues have become subordinated to gender demands defined within the domestic domain. This context would, therefore, limit the symbolic value of these issues for the wider population. The example she gives is the way the work of women journalists would be limited to the 'women's section' of a newspaper if it was specifically delineated as a 'female issue'.

García is of the opinion that ecological movements led by women tend to be more successful at opening new spaces of political significance, that is, transforming the ecological into a new political fact (García Guadilla 1993). However, the assumption is that this can in turn only be possible if the connections, financial assistance and time are available. For these reasons, it is argued, the middle- to upper-class women's ecological movements have been more successful.

García rejects the concept of ecofeminism on the grounds that it is essentialist, and argues that it reinforces socially prescribed female roles. While she recognises that women are the ones that are in control of the non-money economy and the systems of reproduction and survival she equates the issue of ecofeminism with the World Bank's and the United Nation's view. This has been the result of undervaluing knowledge of a special relationship with nature held usually by poor women, due to their close integration with the environment.

RESPONSES – ECOLOGICAL SOCIAL MOVEMENTS

The damage to the environment that has been carried out in the name of development has drawn a response from a number of sources. The active resistance in defence of the ecosystem seems to be a bright spot on the horizon, and proves that if legal rights were extended to the social groups currently most dependent on the sustained existence of the natural environment, the challenge of its protection would be effectively taken up.

The Indigenous people, sometimes spontaneously and sometimes with the support of environmental groups, have for some years been staging protests in defence of their land rights and culture, so belying the traditional Western view of the fatalistic Indian awaiting extinction. Among highland groups, Indians have organised peasant associations, taken up arms and become involved in armed struggles in Guatemala and Peru, while in the lowland areas a plethora of Indian organisations have sprung up, demanding land and help from the government, and defending themselves against the invasions of agribusiness, mining companies and poor peasant colonisers. Mario Juruna, the first Indian to become a deputy in Brazil's parliament, made his people's demand quite clear: 'Indian wealth lies in customs and communal traditions and land which is sacred, Indians can and want to choose their own road, and their road is not civilisation made by whites . . . Indian civilisation is more human. We do not want paternalistic protection from whites. Indians today . . . want political power' (Cited in Green 1991: 166). One of the largest movements was the meeting that took place at Altamira, on the Xingu river in 1989. Nine hundred Indian people (Dorey 1990), many of whom had been warring for years, were united in their protest at the series of dams planned for the Xingu river. Two of the dams to be built near Altamira would involve the flooding of land inhabited by the Kayapo and other tribal peoples. Originally set up by the Kayapo leader, the meeting received a good deal of support from national and international human rights organisations, who brought to the event considerable media attention, including 200 journalists and twenty television crews (Dorey 1990). The Indians added to the spectacle by wearing warpaint and carrying bows, arrows, spears and machetes and from time to time bursting into war songs and dances.

The Indians wanted to know who would be the real beneficiaries of the dam. Most of the constructors of the dam were Western development agencies and banks or were at least financed by foreigners, and the wealth generated largely went to pay off the country's debt (Dorey 1990). No one in Brazil was really benefiting from the project, except perhaps in terms of the prestige it would bring the country. A measure of success was achieved as plans were dropped for some of the dams.

The existence of a coordinating body (COICA) and also a national autonomous organisation, the Union of Indian Nations (UNI), suggests the high level of organisation that tribal groups have reached in their attempts to protect their homeland. The UNI has existed since 1980 and has also supported individual campaigns over land demarcation, as well as representing the country's 220 000 Indians in Congress and in negotiations with government departments. Increasing numbers of Indians

have stood as candidates in municipal, state and federal elections, en-
abling them to develop their experience and understanding of the
political process in Brazil (Treece 1990: 283).

The ability of the Indians to defend the land has been proved by their
participation in the lobbying of the Constituent Assembly in Brazil,
which received a daily presence of more than 200 representatives from
over thirty Indian nations (Treece 1990). The new constitution now
extends the right to vote to Indians, and guarantees their possession of
the land they inhabit, together with their surface resources (MacDonald
1991). These measures, however, are rarely enforced. The UNI is not
officially recognised by the Brazilian government, which cannot accept
the concept of separate Indian nations and rules out the concept of a
multicultural nation.

The UNI also works to redress the balance within communities that
are being strongly influenced by the outside and promote Indian
culture. They provide an Indigenous doctor therefore for the hospital
funded by the Catholic Churches Missionary Organisation (CIMI)
which works according to a belief in Liberation Theology and therefore
to assist the Indigenous people in their own initiatives. This hospital
serves the Kaxinawás, the Kulinas and the Katukina who live along the
Envira river in the westernmost part of the state of Acre, and who have
also joined together in their own organisation which they have called the
Organisation of the Indian Peoples of the River Envira (OPIRE), formed
in 1988.

The UNI has also joined forces with the rubber-tappers, creating the
Alliance of the Forest Peoples. The rubber-tappers have championed
the cause both nationally and internationally of the people who must
defend the forest as they defend themselves. Chico Mendes was their
leader. He was the president of the Xapuri Rural Workers' Union, a
member of the National Council of Rubber-tappers, a member of the
national council of the Trades Union Congress (CUT), an activist in the
Workers' Party (PT) and committed to the defence of the Amazonian
ecosystem. In 1985 he advised the World Bank on Amazonian development projects. In 1987 he received the Global 500 prize for defence of
the ecosystem from the United Nations and a medal from the Society
for a better world. At 5.45 p.m. on Thursday, 22 December 1988, Chico
Mendes, trade unionist, rubber-tapper and ecologist was assassinated in
the doorway of his home in Xapuri, Acre (Mendes 1992: 144).

It was not until the 1970s that the rubber-tappers broke free from the
serfdom of the rubber estate owners. But no sooner had the rubber-
tappers started to live independently than the new ranchers started to
move in and exploit the forest in a way that is destroying it. The new
ranching companies come bringing big money from the south, often
augmented by government incentives. The rubber-tappers find themselves facing an increasingly bitter fight for the forest that feeds them.

The Xapuri Rural Workers' Union has fought by developing the technique of the *empate*. During the dry season ranchers hire labourers to clear the forest for pasture. Just before the rain comes in September the cleared areas are fired. Faced with eviction and loss of livelihood, the rubber-tappers begin to assemble *en masse* at sites about to be cleared, preventing the clearing and persuading the labourers to lay down their chainsaws and go home. Since 1975 the forests of the Upper Acre valley have been the scene of forty-five *empates*, fifteen of which were successful. These led to about 400 arrests, forty cases of torture and some assassinations, but the resistance saved more than 1 200 000 hectares of forest (Mendes and Gross 1989).

In Colombia, in contrast to the colonisation and exploitation taking place in neighbouring Brazil, there is a new policy of protecting the Amazon and its people. In April 1988, deep in the heart of the Colombian Amazon, at La Chorrera, the President of Colombia, Vigilio Barco, told the leaders of the Indian Community: 'I have come to give some good news, a word of truth: at last the land which is yours is yours.' With that he returned 6 million hectares of the Predio Putamayo region to its rightful owners (Bunyard 1989). The policy of returning the lands to the Indigenous inhabitants was promoted by the Director of Indian Affairs, the anthropologist Martin Hildebrand. As a result, 18 million hectares of the Colombian Amazon now belong to some 70 000 Indians of more than fifty ethnic groups, making the region the largest contiguous territory belonging 'legally' to tribal peoples. Since 1968 the government has created some 242 *resguardos* and nineteen reservations, giving land rights over nearly 26 million hectares to 180 000 Indians across the country (Bunyard 1989).

The new governmental position and the creation of *resguardos* has been followed by a remarkable resurgence of the traditional way of life. In the 1950s the number of communal houses in known Indigenous communities was estimated at five. Today there are about forty and rising fast. The chiefs and shamans are also gradually regaining authority, an authority which is recognised by the state itself (Bunyard 1989).

The government expectation is that the return to a traditional way of life will lead to the protection of the rainforest. Bolivia is now seeking advice from the Colombian lawyer on how to create *resguardos* for the Indigenous people of its Amazon region.

Some of the most sophisticated Indigenous organisations of the region inhabit Ecuadorian Amazonia, or the Oriente as it is locally known. The Shuar in southern Oriente were one of the first Indigenous groups in Latin America to become politically organised, achieving some notoriety for establishing their own radio station. In 1990 the last remaining unorganised ethnic group established an organisation and with the others became affiliated to the CONFENIAE (Confederation of the Indigenous Nationalities of the Ecuadorian Amazon), which acts as a regional and national voice for Indigenous issues and politics (Nelson 1992). Hundreds of communities joined the Amazonian Indian uprising of mid 1990, blocking roads, invading haciendas and marching through the provincial capitals. The Indian movement virtually took over the highlands, speaking with a united voice and forcing the government to address the land problem. In early 1992 over 2000 Amazon Indians marched up from the Pastaza forests to the capital and camped out in a central park. Once again the main issue was land. The Amazon groups want a legal title to long-occupied forest lands which are now being infiltrated by new settlers, gold-panning families and oil companies (Kendall 1992).

CONFENIAE focuses primarily on the recognition of traditional land rights identifying its principal objective as 'the realisation of autonomy and self-determination for all Oriente ethnic groups, who see themselves as separate nationalities incorporated unwillingly into Ecuador's political boundaries' (Nelson 1992: 30). Again, CONFENIAE are concerned that the devastation of the Oriente, supposedly for development purposes, does not benefit the mass of people who live there. The major 'development' project has been the extraction of oil, but profits from petroleum leave the region and a significant proportion leaves the country. Tribal groups such as the Cofan who have lost land because of the extraction of oil are left in poverty, with less security than previously and a distinct cultural loss.

It was some 100 000 unexpectedly well-organised Amazonian Indians that were the driving force behind the formation of the Confederation of Indigenous Nationalities of Ecuador (CONAIE) (*Financial Times*, 11 December 1992), which is in turn affiliated to COICA. In spite of their success in drawing public attention, the Indians are not considering taking on the political establishment as yet. The differences between the organisations are not insuperable, said one Shuar Indian, but 'it is too soon for a political party. We have to go slowly and learn how to think ahead and how to work with each other' (*Financial Times*, 11 December 1992: 36).

Direct action by Indigenous groups in defence of their land has a long history in Southern Mexico, and the incidence of which has by no means declined. In fact it was commonplace in the 1970s as local people regularly blocked roads and destroyed machinery. Defence was mainly directed against loggers, settlers and ranchers in this area where there are two million hectares of tropical forest. In July 1986 the logging in Chimalapas gained national publicity when representatives of the Chimalapas communities met with the governor of Oaxaca to protest against logging by, amongst others, the brother of the governor of the neighbouring state of Chiapas, but no action was taken. In December 1986 local activists ('*chimas*') 'kidnapped' ten men, whom they accused of being illegal loggers, land invaders, drug-traffickers and cattle thieves, and among whom were the governor's brother, his nephew and four gunmen. The *chimas* burned the men's centre of operation in the forest, confiscated or destroyed machinery and took the men to the municipal office in Santa María de Chimalapas where they held them as security against their demands. These included compensation for illegal logging of over 80 000 hectares of forest and the termination of logging licences. Government representatives, police, helicopters and army units descended on the village where they quickly agreed to implement a 1967 presidential decree recognising Indian land rights to over 460 000 hectares of communal land on the state boundary between Oaxaca and Chiapas (Blauert and Guidi 1992).

After the 1986 'kidnapping', the communities of the Chimalapas continued to negotiate with various authorities, publicise their situation and campaign in alliance with environmental organisations. They want the establishment of a locally defined and controlled biosphere reserve, bringing together some thirty *ejidos* (communal land holdings resulting from agrarian reform) and fifty villages in the 250 000-hectare area. The most important thing about such a 'protective productive area' is control over local resources in a way that ensures not only environmental sustainability but also political control and economic development according to locally defined criteria (Blauert and Guidi 1992).

There is always the risk that organisations such as these may be co-opted by the government or create a new Indigenous élite. However, their commitment to a bottom-up structure should go a long way in preventing this from happening.

RESPONSES – WOMEN'S ENVIRONMENTAL MOVEMENTS

Many case studies of female-led organisations fighting to protect the environment can be found in Africa and India. The link between women in these countries to the natural resource base is more direct than in Latin America. The women are the water-carriers, the wood-collectors and the field-workers by prescription of many of the cultures of these areas. In Latin America the woman's role is slightly less formularised and for this reason her work less recognised.

However, national and local women's groups can also be found in Latin America. Maria Pilar García Guadilla is one writer who has devoted some study to responses to the ecological crisis in Venezuela. She dates the movement back to the 1930s with the pioneering scientific conservation societies.

As well as in Brazil, there has been a rising incidence of social movements in Venezuela. More than a hundred environmental organisations now exist. The economic crisis of the 1980s threatened many people's life-styles and 'given its identification with poverty, environmentalisation of urban demands has occurred particularly in the 1980s' (García Guadilla 1993: 71). But it is also due to the increasingly authoritarian nature of the civilian government. Venezuela has had a formal democratic government since 1958; however, the constitution institutionalised in 1961 had strong paternalist, populist and clientelist features sacrificing representational democracy for political stability. It is with the economic crisis of the 1980s, though, that the formal political structure has become delegitimised and people have sought new spaces to 'create shared meanings about a desirable society' (García Guadilla 1993: 69).

As in Mexico and Brazil, these movements are organised according to many different points of view (Quadri 1990; Puig 1990). García finds six different types of organisations in Venezuela using a method which divides the group according to sex composition, socioeconomic, ideological and cultural characteristics. These include political ideological organisations which are in the main attended and led by men (they normally have an ambivalent attitude towards gender-specific issues and prefer not to divide the organisations); symbolic cultural organisations, organised by middle- and upper-class women; and neighbourhood associations – usually led by men but the bulk of active members being women.

Symbolic cultural movements have such a title because they point to problems the solution of which requires the 'decoding of dominant models and searching for alternative meanings and orientations for political actions in the cultural sphere' (García Guadilla 1993: 74).

AMIGRANSA, founded by five women in 1985 (who now make up the executive committee of the non-profit-making civil association), is an example cited by García (1992) of a successful cultural symbolic movement. They have been largely successful in their demands, the main one of which concerned the defence of the Canaima National Park. Successful mobilisation of the group occurred when the country was threatened with a 'rally *transamazónico*' (car drivers' competition which would have ended in Brazil). Their success in putting a stop to such projects apparently lies in the successful manipulation of the mass media, whereby a new political space was opened and the 'environment' was billed as a new political phenomenon. García cites various other female-led neighbourhood groups concerned with the environment.

Women's organisations also proliferate in other parts of Latin America. The Friends of the Earth (Acao Democrática Feminina Gaúcha: ADFG) of Brazil was founded in 1964 as a women's organisation. In 1983 it became possible for men to become associates, although women's issues remain central. In 1974 environmental protection and the promotion of sustainable development was added to ADFG's agenda (María Guazzelli, ADFG, Brazil, cited in Dankelman and Davidson 1988: 149).

Another example of what can be done to improve the situation of women in agriculture by reclaiming the agriculture that will sustain women, their families and the environment for years to come, is evident in the mountains to the south of the town of San Cristobal de las Casas in Chiapas, Mexico, where Totzil Indian peasant families live.

There is in this region a strong tradition of collectivity: and water supplies, electricity points and children's schools have all been established. Yet during this process the story for the women sounded a familiar note. 'However crucial the women were to the success of the projects, women were allowed no part in the structure or decision making of local organisations' (Dankelman and Davidson 1988: 25). The cooperatives of most of the communities were male-dominated and did not give priority to the activities for which women were responsible, such as the rearing of small animals. Women also suffered discrimination from a local revolving fund, which allowed them no access to loans for improving their economic position.

The response of the women in the village of Pinabetal was to set up their own collective to tackle their gender role problems such as lack of access to land, no roads, no schools and no work. Thirty peasant women are now involved in the Pinabetal women's collective. They grow vegetables, raise sheep for wool and a source of natural fertiliser, and are planning to keep dairy cattle to provide milk. OXFAM also supports projects such as chicken rearing, water management, more vegetable plots and soil conservation measures. Other activities include study groups, evaluation and planning sessions and discussions about Indigenous women farmers and how best to spread the word to help other women to find ways of working together to solve the problems.

Ethnicities and gender come together in some cases such as the women leaders of Black communities from the Calima river, in Valle de Cauca, who have set up a coordinating committee of all communities on the river. They see pressure for immediate black land rights as crucial in their situation on the crossroads of North Central and South America, which has over the past decade become the target of the ambitious Plan Pacífico. This would not only exploit the resources of the area but also look to promote it as a trade platform. The Black and Indigenous people have formed dozens of community organisations in this area to protect land rights (Barnes 1993).

CONCLUSION

Notions of development are again put under the microscope when addressed in the context of the environment. It is precisely the drive for so-called development in the form of rapid economic growth that has produced a number of contemporary urban and rural environmental problems. Industrialisation, seen as the key to growth, has been driven forward regardless of environmental safeguards. The need for energy to fuel this growth has taken precedence over the livelihoods of the indigenous people. Involvement in a global economy has made possible the financing of technological installations which have caused pollution and created the demand for exports which drain the land of its long-term value. The characteristic inequality of the region produces a pattern of land-holding which requires the poor to resort to ecologically damaging practices to survive. The concepts of sustainable development and sustainable livelihoods suggest a new style of development is appropriate.

A variety of solutions, including 'top–down' and 'bottom–up', have been entertained. It is the contention of this chapter that those which focus on the participants as subjects have more potential for producing sustainable livelihoods in the future. The collective response of those involved has been strengthened by the coincidence of ethnic and environmental concerns. The need to maintain landholdings to survive reinforces the desire of Indigenous people to preserve traditional land rights, and along with that, ethnic identity. Ecological issues have brought a raising of consciousness, expressed in the growth of ethnic social movements, some of which bring together a number of different tribes, but united in a sense of nationhood.

FURTHER READING

Adams, W. (1990) *Green development, environment and sustainability in the Third World*. Routledge, London.

Cockburn, A. and Hecht, S. (1989) *The fate of the forest: Developers, destroyers and defenders of the Amazon*. Verso, London.

Dankelman, I. and Davidson, J. (1988) *Women and the environment in the Third World, alliance for the future*, Earthscan publications Ltd, London.

Green, D. (1991) *Faces of Latin America*. LAB, London.

Goodman, D., Redclift, M. (1991) *Environment and development in Latin America: The politics of sustainability*. Manchester University Press, Manchester.

Redclift, M. (1987) *Sustainable development, exploring the contradictions*. Methuen, London and New York.

Treece, D. (1987) *Bound in misery and iron: The impact of Grande Carajás programme on the Indians of Brazil*. A report from Survival International, London.

4

Values and institutions

Hispanic heritage was a major element contributing to the cultural formation of Latin America and, since values constitute a vital component of culture, the Hispanic influence has been considerable in shaping norms. At the heart of these have been the institutions of church and family and a code of behaviour based on personal honour. Commentators on Spanish culture through the centuries have pointed to its emphasis on dignity and the corresponding fear of shame which leads to behaviour that is continually concerned to respect others and, above all, their individuality (Pitt-Rivers 1977). The resulting emphasis on personalism and personalised relationships has become very much a feature of social life in the New World, too, as will be seen in the next chapter.

CHURCH

Durkheim's investigations into religion, with their historic results, were prompted by the fact that he saw religion as the major source of values and, despite increasing secularisation, religion still plays an influential role in society's moral codes. With the exception of particular indigenous groups of Latin America and some communities who have been evangelised by Protestant missionaries, Latin Americans are Roman Catholic. The conquest brought not only soldiers to Latin America but also priests and missionaries, for the Christianising mission of the Spaniards and the Portuguese was vital to their cause. These priests were instructed to, if possible, incorporate and then translate pagan religion into the Catholic orthodoxy, rather than destroying completely these alien beliefs. This proved more feasible for the Incas than the Aztecs, but eventually, with varying degrees of repression, the Indigenous people were brought into the Catholic fold. Some historians have argued that, once the Indians had become part of the Christian community, their cause was championed by the church, which protected them from some of the worst excesses of colonial power (Poblete 1970). It must be remembered, however, that although priests may have alleviated the lot of a small number, the colonial epoch still retains a place in history which is notorious for the brutality shown to the Indians.

During the colonial period the church was closely aligned with the state, giving full support to the dominant power and, in return, receiving special privileges. The church for its part acted as an administrative agency of colonial expansion and a major institution of social control. After political independence, however, many Latin American countries became involved in conservative–liberal struggles, forcing the church to make a choice between élite groups. In general the church allied itself with the conservative sector, knowing it could then rely on the wealth and status of this group, and in return, offered a value system that tended to supported conservative regimes. It was in this role that the Catholic church entered the twentieth century. 'In the main, the church and its leaders drew their importance from the support they gave to the existing powers and from their multiple involvements in education, social welfare and administration' (Vallier 1967: 193).

The church, having survived independence movements, intra-élite struggles and, in the case of Mexico, the virulent anti-clericalism of the revolution, continues to be an important institution in twentieth-century Latin America. Its strength lies partly in its scope, for Roman Catholics make up about 80 per cent of the population; for instance, the figure is 92 per cent for Argentina and 80 per cent for Chile (World in Figures 1987) and also in its institutional links with other aspects of life such as education, health, leisure and community activities. But the power and attitudes of the church vary considerably within different Latin American countries. Colombia, often considered the most traditional of the Latin American republics, is a clerical country. The church has a strong hold on education, the clergy command great respect from the population and the church leaders, like other Colombian élites, have not favoured social change. The Chilean church, however, has often been referred to as 'the most progressive Catholic system in Latin America' (Vallier 1967: 220). The hierarchy, often in the hands of the liberals, has promoted reform programmes, such as technical training for peasants and the forming of cooperatives in the shanty towns, has become involved in politics through the reformist Christian Democrat Party and has opposed the brutalities of the military regime since 1973 by speaking out openly against its excesses and providing soup kitchens for the needy. Mexico is a country where the church does not have a strong grip on the population because of the force with which the revolutionaries opposed the clergy, believing them to be partly responsible for the backwardness of rural areas. The revolution made education secular and reduced the wealth of the church by confiscating some of its land. Today, although the Catholic church has been reinstated and crucifixes and images of the Virgin Mary and the saints are to be found adorning many homes (especially in rural areas), the church's influence on the dominant groups in society is not great. Thus what is at stake is not so much the continuing vigour of the church but the direction which it will take. Will it continue to uphold the conservative values of those in power, or will it offer a more radical alternative?

In this century differences have emerged between leading churchmen in the region and, especially since the 1960s, these differences have developed into significant splits within the church. Prior to the middle of this century, the Catholic church had concerned itself predominantly with spiritual matters. The Second Vatican Council (1962–65), however, redirected the church's interest towards the temporal world and especially issues such as justice, human rights and freedom. Not only was the traditional ideal of union between church and state relinquished but the Council also stressed the responsibility of the church to pass moral judgements, even on matters touching the political order, whenever basic personal rights or the salvation of souls make such judgements necessary. 'Many bishops in Latin America, Asia, and Africa, since the end of the Council, have exercised this prophetic role vigorously, denouncing political disappearances, torture, economic exploitation, and racism perpetrated by authoritarian regimes in their countries' (Smith 1982: 5).

This shift in official church policy has made possible a greater integration of religious and secular values and also the move from legitimisation of the existing social structures towards the promotion of greater equality.

In Latin America the Second Vatican Council was followed up by a meeting of bishops from all over the continent in Medellín, Colombia. The message to emerge from this conference was one of liberation, from both spiritual and material bondage, and this liberation was to be achieved by the participation in the whole life of the church, not just in church services. The conference also stressed the development of Christian Base Communities (CEBs). This is an aspect of pastoral development in which small groups meet together principally for bible study but, in many cases, they may be devoted to promoting other activities as well, such as literacy classes or agricultural cooperatives. The CEBs represent a more radical move for the church for various reasons. They are, for a start, a product of the local community, not the church hierarchy. They are organised and led by the people and although a religious leader is present to offer assistance, the nature and activities of the group are decided by the members. The organisation is highly democratic, for lay readers are elected and then must consult their members in decision-making. Many nuns and priests who have worked with the CEBs have found that members become much more self-confident and are less fatalistic in their attitudes (Montgomery 1983).

Accompanying these changes in the policy and organisation of the church were the new developments in theology, which have come to be known as the Theology of Liberation. This is a truly radical development for the church, because it takes much of its thinking from Marxism. Liberation Theology explicitly links society, politics and religion and in so doing, bases much of its social analysis on dependency theory. Thus poverty, injustice and other secular problems are seen as being the result of exploitation by world capitalist powers. Any sort of improvement for the masses must involve a reordering of the social structure. This change cannot take place without the involvement of the people themselves and, therefore, they must be moved away from their traditional passivity towards an active role in shaping their own lives. In line with the movement towards participation in temporal affairs, liberationists believe that they must demonstrate genuine Christian love by committing the church to creating a more egalitarian society through religiously inspired political action. Liberation Theology therefore advocates action and exhorts the clergy to emerge from their cloisters to help the people to help themselves.

The call for action has been put into practice in some parts of Latin America, especially Central America. After the successful Sandinista revolution in Nicaragua, two priests were included in the Marxist government. This was going too far for the established church both in Rome and Nicaragua, with the result that the Pope called for their resignations. In 1983 the Pope, concerned by the spread of Liberation Theology, visited Central America, ostensibly in order to promote church unity. The Nicaraguan part of the tour was marked with a well-publicised clash between the Pope and Sandinistas, which was prompted by a briefing-paper presented to the Pope by church right-wingers.

The same split between conservatives and progressives can also be found in Guatemala, Honduras and El Salvador. In the latter, the progressives have been active in the top echelons of the church hierarchy. Archbishop Oscar Romero, although thought to be a moderate when he took up the Archbishopric, proved himself to be a zealous proponent of Liberation Theology, broadcasting his masses, which included political messages, to the people throughout the country by radio. Romero linked his readings with the reality of life in El Salvador and each week his sermon was followed by 'a reading of every documented case of persons who had been killed, assaulted, tortured or disappeared by any group on the left or right' (Montgomery 1983: 78). Since far more attacks were brought about by the government's death squads than the Left, Romero's vast audience soon became aware of what was happening in their country. Romero's beliefs and bravery, however, cost him his life for on 24 March 1980 he too was assassinated by a professional hit-man hired by the extreme Right.

These events in Central America demonstrate the church at its most radical. Many churchmen, however, have not been happy with these developments, which have given rise to schisms threatening enough to warrant the Pope's visit. Some church leaders fear that a political commitment on the part of the church would undermine the value of the spiritual dimensions of the Christian faith, for it threatens their mission of salvation. For many, the whole idea of using Marxist concepts of analysis is out of the question for a Christian, and the spectre of communism has been one of the most divisive aspects. Another fear of the critics of Liberation Theology is that the logic of the argument leads to violence, since violence may be the only means available to overthrow corrupt regimes, and this they feel no Christian should condone.

In many areas the split is by no means a clear-cut one of those for and against, but often includes middle groups who are sympathetic to some of the ideas of Liberation Theology but who cannot wholly agree with it. In Central America, three groupings can be identified. The institutional church, preoccupied with orthodoxy and fundamentally opposed to Marxism because it is viewed as materialistic and atheistic, is made up of bishops, priests, laity and religious movements who, though not very numerous, are powerful, partly because the laity of the group belong to the wealthier classes. The centrist sector, who favour a middle way, is opposed to repression and violence whether it is from the Right or the Left. This group, which does not ally itself with either American intervention or Marxists, represents the greater part of the church in Central America. The popular church, with its base in the CEBs and animated by Liberation Theology, feels that risks must be taken to right the social and political evils of the world, and so is sympathetic to revolutionary movements. Though not large, this sector of the church is qualitatively important.

Max Weber suggested that the protestant ethic produced the value orientations and attitudes in which capitalism could prosper, and that Catholicism, like most other religions, hinders social progress. It could be argued that Latin America offers a good example of this. For centuries, Catholicism allied itself with conservative forces opposed to progress, but now that elements within the church have come out in favour of change it appears to be opting for a more socialist rather than capitalist line. 'Traditional Catholic principles, once an obstacle to change in emerging capitalist economies, when they were identified with corporatist ideologies, now appear to be more compatible with various socialist models of development' (Smith 1982: 4).

In fact it could be argued that the shift in Catholic theology came at a time, after the Cuban revolution, when the spectre of the triumph of socialism actually seemed a real possibility. It follows, therefore, that this was, possibly, an attempt to maintain a congregation in a situation that would normally be hostile to the very principle of religion.

With the international crisis of confidence in the socialist model, however, the Catholic church of the 1990s fears losing its flock to totally different forces. Membership of US-based born again protestant groups has snowballed in Latin America since the late 1960s to such an extent that it has led to comments suggesting that Latin America is in fact 'ripe for reformation' (Green 1991: 182). At the beginning of the 1970s evangelical protestants counted some five million members, excluding children. Two decades later that constituency extended to some forty million, which is remarkable even allowing for rapid growth in population (Green 1991: 182). It may even be the case that in areas of Latin America, such as parts of Guatemala, the numbers of protestants regularly involved in worship and fellowship exceeds the number of Catholics (Martin 1990: 50). The deepest penetrations have occurred in Brazil, Chile, Nicaragua and Guatemala – the areas of greatest poverty and strife. According to the *New York Times* for 25 October 1987, evangelicals in Brazil have doubled in number since 1980 to about twelve million, and in 1985 there were 15 000 full-time protestant pastors in Brazil compared to 13 176 priests (Martin 1990: 51). 'The most dramatic changes are in Central America. Guatemala is now 30 per cent protestant, Nicaragua 20 per cent and Costa Rica not far behind with 16 per cent' (Martin 1990: 51).

The protestant movement is characterised by fragmentation and numerous community-based sects are taking root. Usually rabidly anticommunist in its attitude, the movement is promoted through the evangelist capacity to unite political conservatism and modern technology. Tele-evangelists from the USA buy thousands of hours of air time on national television stations. In Rio de Janeiro evangelicals control twenty-five of the thirty-three radio stations, forcing bishops to take to the airways themselves in self-defence (*Financial Times*, 2 October 1993: II/X). Pentecostalism, the most dynamic strain of evangelism, has become the fastest growing denomination in Latin America. Pentecostals have been the first popular manifestations of protestantism and constitute an extensive engagement with the poor. This sect stresses the gifts of the spirit and maintains that true salvation lies in one's individual, personal, relationship with God. Speaking in tongues and faith-healing form part of services. As it offers the attraction of spirituality as well as a supportive community and a sense of purpose and identity, it flourishes in times of war or in new urban situations in the face of isolation or confusion. For Liberation Theologians, it presents a powerful force of regression back to passivity before the dynamics of power and state (Martin 1990: 54).

EDUCATION

Other institutions have come to challenge the church's traditional dominance of the value system. As mentioned in Chapter 1, education has an important part to play in development, both in its contribution to structural change and to changing values. In the former, education can provide more people with the skills and information to secure them good jobs and thus produce an increase in the number of people prepared for fulfilling and materially rewarding occupations. This can open up new opportunities and create social mobility. Values are maintained or altered because education is one of society's mechanisms of socialisation. Education, too, can be directed to encourage certain ideals.

Education is a sector which has expanded greatly in Latin America. There is no doubt that it is a growth area, which has had distinct achievements in many countries. Traditionally, the prerogative of the élite, education is now available to most Latin Americans.

> Although no generalisations can be made about the behaviour of all education systems in Latin America some tendencies can be observed. As a rule urban schools are better resourced than those in rural areas. Schools for the high and medium income brackets are well equipped, while those for the poorer classes are badly provided for. In mixed societies, from the ethnic point of view the 'white' schools do better than those for children of mixed or 'pure' ethnic origins. According to statistical information some countries are in general terms doing better than others. Argentina, Brazil, Chile, Cuba, Mexico, Guyana and Peru appear to be doing better than countries like Bolivia, Ecuador, Haiti, El Salvador and Guatemala, while countries like Venezuela and Costa Rica seem to be in between those types of countries.
> (Albornoz 1993: 6)

Have these developments brought about social mobility? The difficulty with the argument that more and better education leads to mobility, is that better-trained people do not, on their own, lead to an increase in the number of better-paid jobs. As Kahl has said, education 'reinforces the existing status system and provides for movement within it' (Kahl 1970), but this suggests that it does not change the system. This, however, still overlooks the fact that large numbers of the better educated have no power base from which to bring about change if there are no jobs or positions in society for them. What has been happening in Latin America is that the growth of education has been much faster than structural change in society, with the result that some of the potential value of education for social mobility is lost. Some Latin American educationalists predict that the ensuing frustration of an educated population unable to find appropriate work will lead to social conflict.

In quantitative and qualitative terms, how extensive have the changes in education been? Literacy rates, a good indicator of educational achievement, have improved very considerably during the century. In

1895 the literacy rate for the population aged fourteen years and over was 53 per cent in Argentina, and 68 per cent in Chile. Even by the second decade of this century, the illiteracy rate for the Brazilian population was about 80 per cent. By 1950 there had been improvements, but they were slow in taking place. In that year, with the exception of Argentina, Chile, Cuba and Uruguay, the illiteracy rate of populations aged over fifteen years, exceeded 30 per cent in all Latin American countries, 50 per cent being recorded in Brazil and higher percentages in Central America (Rama 1983: 16). It is since the 1950s that the expansion in education has really accelerated, with a resulting boost to literacy rates. Now Argentina, Chile, Uruguay, Paraguay, Cuba, Costa Rica and even Belize have all topped the 90 per cent mark, with 95, 93, 96, 90, 95, 93, 93 per cent (1990) respectively. Even Mexico and Peru are not far off, with 87 per cent and 85 per cent respectively (1990). Only for Haiti is the literacy rate particularly low with 52 per cent in 1990 (Third World Guide 1992: 39).

These improved rates have been brought about by increasing school enrolment. In the period from 1950–80, the school age population (under twenty-four and over five) which was attending school doubled. By 1980 in Argentina six out of every ten individuals of school age were incorporated in the educational system. Much of the growth took place in the 1970s, when it was concentrated more on the higher levels of the educational system. Between 1950 and 1980, participation in the higher levels of education rose from 5 to 16 per cent, in secondary from 15 to 25 per cent, and primary education from 50 to 90 per cent (Filgueira 1983: 69). Education in recent years has been characterised by less rigidity, less concentration and greater opportunities for mobility, but all these features become more apparent higher up the educational pyramid. These changes have brought about a reduction in inequality in education throughout the region.

This expansion of education has not brought about a similar increase in social mobility. Class background is an important factor in the level of schooling achieved, partly because the higher the level the greater the cost (if only in terms of opportunity costs) and partly because the increased educational facilities are best in the well-to-do residential areas. Eckstein found in Mexico that in over three generations there had been little occupational mobility, especially into the white-collar class (Eckstein 1977).

But it is not just that access to the increased educational facilities is conditioned by social background; even where people achieve educational mobility, the economic structure has not provided a corresponding increase in jobs to match the mobility, especially in occupational terms. Eckstein found that the best educated had access to the most rewarding jobs, 'but because such jobs are not expanding as rapidly as schooling, education in itself provides no guaranteed job' (Eckstein 1977: 205). Filguera points out that the reduction of inequality in education is in contrast to the behaviour of income distribution, where inequity has been little reduced during the same periods, and which shows extreme relative rigidity and concentration in comparison with the greater equalizing capacity of the educational system. The figures for inequality in income distribution in the region (for example, Uruguay, 1.49; Brazil, 0.70; Chile, 0.5) are always higher than those for educational inequality, and dynamic trends suggest that the gap between them tends only to widen.

(Filguera 1983: 70)

Filguera suggests that since the major educational developments have taken place so recently it may be too soon for them to have brought a greater reorganisation in society. He does add, however, that structural changes are also necessary to accommodate the benefits of educational expansion.

The growth of formal education systems could not have taken place without the incorporation of certain sectors of society which had previously been excluded. The expansion of primary education has benefited the peasantry and the urban poor in particular, since rural education has lagged behind urban education. A United Nations study showed that most of the students completing secondary school represented the first generation of their family to reach this level (Filguera 1983: 71). Those who reap the benefits of higher education, however, are recruited predominantly from the middle classes. A study of Uruguayan professionals who graduated in 1970 revealed that barely 11 per cent came from families in which the father worked in skilled or semi-skilled manual occupations. Mobility, however, is greater if educational levels of fathers and graduates is measured. Eighteen per cent of the professionals had fathers with incomplete primary schooling and 27 per cent came from families where the father had achieved no more than a primary school education. The graduates had frequently been more successful than their fathers in educational achievement but, in general, were entering occupations of similar status to that of their fathers (Filguera 1983).

The proportion of students from poorer families who enter higher education is similar to other Latin American countries, though there are variations according to courses and institutions. Short and intermediate courses, and those which do not involve the student in heavy costs tend to have a more democratic social composition. But the greater the prestige and reputation of the institution, the more it will recruit from the upper echelons of society.

A university education confers prestige and reputation, thus reinforcing the enormous gap between the privileged group and the masses. What is intended as a democratic institution, and indeed fees at many state universities are surprisingly low, in fact ends up being highly élitest. Academic entry to universities may be facilitated by colleges offering preparatory courses, with the result that some universities in the region recruit very high numbers. The National Autonomous University of

Mexico (UNAM) in Mexico City brings 400 000 students onto a campus big enough to warrant two metro stations, its own road circuit and sports facilities of Olympic standard. Drop-out rates, however, are often high because students can not afford to support themselves through the lengthy degree courses. Educational reforms, far from opening up either society or education to social mobility, have the function of reproducing the social organisation of society and acting as an expression of the political system. They therefore become instruments for social stability.

It has been argued that despite the great variety of institutions of higher education (private/public, traditional/modern, denominational/non-denominational and metropolitan/non-metropolitan), with a few exceptions in Mexico and Brazil, the universities do not achieve high academic levels. By focusing on a bureaucratic and political culture they fulfil social rather than intellectual demands. Social selection tends to be favoured rather than intellectual and, as the main function is considered to be teaching rather than research, the emphasis is on training professionals and well-qualified experts to fill the ranks of the élite, rather than producing academics (Albornoz 1993).

Although biased toward the higher levels and urban areas, the quantitative development of education has been very significant. However, so far, it seems that this has had little influence on the redistribution of wealth and power in society. This suggests that either educational reform has just not gone far enough or that although improved educational standards constitute an important development goal, they can only be effective if accompanied by structural changes in the rest of society. Perhaps both are the case.

In terms of values, education is the way in which young people experience socialisation. Traditionally, literacy campaigns in peripheral rural areas would rely on the Bible, which was not only a source of reading material but also a way of encouraging conformity. In the same way, the literacy campaigns of the Cubans and Nicaraguans have used Marxist–Leninist tracts, and other material in line with revolutionary ideas. In the years following the violence of the Mexican revolution great attention was paid to the content of education, in an attempt to involve the young in the principles of the revolution, but this emphasis has declined in the second half of the century.

In general, education has reinforced society's values, but there has been some experimental attempt at providing education to radicalise values. One of the most famous was Freire's Movement of Popular Culture, aimed at raising the consciousness of his students, in Brazil's north-east. Freire wanted to bring people to a critical awareness of their real situation in the socioeconomic context, for only then, he argued, could they really learn and understand. He maintained that the lack of awareness of the needy in Brazil's poorest region facilitated their exploitation. Education, he believed, should offer a means of escaping from the alienation of poverty that bedevils that area (Freire 1968). Freire's experiments were cut short by a military coup, so it is difficult to judge the results. Many schools, colleges and universities in Latin America are part of the church's empire and so the values inculcated are in line with dominant Catholic Hispanic mores. Dissent has occurred at times among university students in attempts to radicalise dominant ideas. One of the best-known occasions was the student movement in Mexico in 1968, when marches, demonstrations and general opposition to the government of the day prompted the intervention of the army. In the final analysis, students, although able to generate new ideas and disseminate them amongst a limited circle of people, do not have power and are no match for the military.

HEALTH

The provision of health and welfare services is another area of recent growth in Latin America which on the surface suggests development, but as with education and resources in general distribution has been very uneven. Throughout the region there has been a great expansion of the hospital building programme and of the training of doctors and nurses. About a quarter of external finance (mostly from the World Bank, the Inter-American Development Bank and the European Community) earmarked for development purposes is allocated to health, nutrition, population, water supply and sanitation projects. In general terms, improvements have been recorded in the indicators for immunisation coverage, maternal and child care, diarrhoeal disease control, actions in health education, detection and treatment of chronic and acute disease, and care in cases of accidents. Results can be seen in statistics: life expectancy at birth has risen from fifty years in 1950–55 (Iturralde 1987) to sixty-eight years in 1990–95 (World Resources 1993).

Yet there are still in some areas immense problems of malnutrition, infant mortality and the spread of diseases which can be easily cured in the industrialised world. In the countries which still have high infant mortality rates, the main causes are invariably linked to malnutrition and the techniques for their prevention and cure are known. In Guatemala, the three major causes of death result from curable diseases: influenza, pneumonia and measles. Cholera, which had been wiped out in the Western world for half a century, reappeared in Peru in the 1990s and spread through Latin America as far as Mexico. It affected countries such as Venezuela, which is regarded as being as modern in parts as some countries of the industrialised world.

The provision of health services is in general multi-institutional, though four categories can be identified:

93

1. Countries with national health systems with provision of services exclusively or primarily under the Ministry of Health, with no significant participation of the private sector, for example, Cuba and Nicaragua.
2. Countries where health care depends mainly on some sort of insurance system, whether private, state or joint. Establishments providing services are owned by the social security system, the Ministry of Health or the private sector. This group includes Argentina, Brazil, Chile, Costa Rica, Haiti, Mexico, Panama and Venezuela.
3. Countries where care is provided mainly through the Ministry of Health, with financing from public funds, together with participation by the social security system, for example, Belize, Bolivia, Colombia, Dominican Republic, Ecuador, El Salvador, Guatemala, Honduras, Paraguay, Peru.
4. Countries where the Ministry of Health and the social security system play almost equal parts in providing health care. The only one of this type is Uruguay.

In almost all of these countries private practice exists, though reliable statistics of its role in the provision of services are hard to find (Iturralde 1987).

How can improvements in health care on an impressive scale be reconciled with the malnutrition and child mortality mentioned in the first chapter and the resurgence of diseases such as cholera? Iturralde suggests that one of the major problems is the absence of well-defined policies concerning research programmes, exacerbated by poor communications between different parts of the various bureaucracies involved. All countries complain of a shortage of financial resources, particularly since the early 1980s with the debt problems that governments have had to face. There is too an emphasis on curative rather than preventative medicine, which explains the high incidence in some areas of diseases which present no threat to health in the industrialised countries.

But a key factor here lies in the distribution of health resources, with the poor frequently lacking access to health services because of economic, cultural and institutional factors. According to Navarro, it parallels the distribution of resources in other areas (Navarro 1976):

1. Dependency – historically due to the export economy, natural resources have gone abroad, similarly in medicine resources leave the region. In this case it is doctors trained in Latin America who are lured by higher salaries and better research opportunities to the USA. The 1970s in particular saw a flood of foreign-trained doctors, about a third of whom were Latin American, to the USA. In the Dominican Republic, approximately half of newborn children die before reaching the age of five and yet 30 per cent of medical school graduates each year emigrate to North America.

2. Distribution within the region reflects distribution of wealth in the society with consumption of health services being urban and primarily in the capital. An analysis of the different sectors of health provision demonstrates the unequal pattern of distribution. Contributions to the private and social security sectors come from 25 per cent of the population, but this sector consumes 60 per cent of all health expenditures. The public sector representing three-quarters of the population consumes 40 per cent of all expenditures. Several countries have severely deprived populations whose situation is not reflected in the national average. In the 1970–80 decade in Brazil life expectancy in the central north-east was forty-nine years and 67.8 in the south of the country, a staggering difference of 18.8 years (Wood and Carvalho 1988). In Guatemala 80 per cent of medical services are concentrated in the capital (Painter 1987).

3. The distribution of medical specialisms reflects a pattern appropriate for the industrialised world rather than the LDCs. Surgery represents the top speciality, with paediatrics and public health being in the lowest categories. In countries where almost half the population is under fifteen years old and morbidity mainly caused by environmental and nutritional deficiencies, this reflects an inappropriate balance of specialisms. These are determined mainly by market forces, with the wealthy demanding the latest cures and operations.

The pattern of specialist distribution emphasises the important distinction between preventative and curative medicine. Nicaragua saw a distinct shift from primarily curative urban-based care for a privileged minority to an emphasis on prevention after 1979. For example children were inoculated against polio and measles with the result that in the following two years, not a single case of polio was reported and measles dropped from being the fifth most common infectious disease to the thirteenth (Painter 1987). Preventative medicine is often cheaper, as in the case of inoculations, which gives it an added advantage for the LDCs.

A country which has made great advances in health care in the region through redistribution is Cuba. Infant mortality has been reduced to thirteen deaths per thousand live births and average life expectancy at birth now exceeds seventy-five years (World Resources 1993). In 1950, 30–40 per cent of Cuba's population and 60 per cent of its rural population were undernourished, but by 1979 malnutrition had been lowered to 5 per cent. Interestingly the national average consumption of calories did not change greatly during that period, demonstrating how averages can conceal wide disparities. By 1979 no one starved in Cuba, nor did any minority eat considerably better than the others (Forster and Handelman 1985). The health system has been reorganised into one based on family practitioners, supported by a network of poly-clinics which provide consultation, teaching and laboratory services. As doctors are paid modest

95

salaries, labour costs are not very high. Achievements have also been reached in medical research, particularly in the areas of developing a new vaccine against meningitis, testing for congenital disorders such as Downs Syndrome and pioneering techniques for treatment of skin diseases.

Significant progress has also been made in the development of social security programmes in Latin America, though again it is not the most needy who benefit. Social Security in the region covers about 40 per cent of the population, though in the majority of countries the proportion is less than 25 per cent (Mesa-Lago 1986). Social Security involves five different programmes: occupational hazards; disability and pensions; health care and money benefits in the case of sickness and maternity; family allowances and unemployment benefits. Of these, pensions form the most common programme in the region. Most Latin American countries are between the stage of social insurance and the more advanced stage of full Social Security. In the majority of cases these programmes are financed by contributions from the individual concerned, his employer and the state. This means that to qualify, beneficiaries must have regular full-time employment. As almost half the population of most big cities are excluded from this modern sector and are compelled to look for more irregular and insecure forms of work, these urban poor fall outside any Social Security provision that may be available. As a result, 'the most needy groups are not provided with social security protection in the vast majority of countries' (Mesa-Lago 1986: 140).

MASS MEDIA

One of the visible contradictions of the Third World that is apparent in Latin America is the spread of sophisticated and technological forms of mass media in poverty-stricken areas. The recently arrived visitor to Mexico City is quick to notice the mass of television aerials that rise from the shacks in the poorest areas of the city. The stereotype of the Mexican peasant, dwarfed by his sombrero and sitting by a cactus should, perhaps, be more realistically represented as the peasant listening to his radio. A study of peasant families in the Cochabamba area of Bolivia revealed that 90 per cent of them owned radios (Ortega 1982). The contradiction is probably the most apparent in Mexico, where cheap radios, televisions and films can easily be imported from the USA, but the media is also very much a feature of contemporary life in other Latin American countries.

Not surprisingly, Latin Americans tend to own more radios than televisions. In terms of statistics, Argentina ranks 44th and Mexico 39th in radio ownership in the world, but when it comes to television ownership they rank 57th and 81st respectively (World in Figures 1987). Television ownership, however, increased rapidly during the 1980s, despite the increased levels of poverty caused by the economic crisis early in the decade (see Tables 4.1 and 4.2). Perhaps at times of rising poverty televisions bought cheaply second-hand or through hire purchase agreements represent one of the most accessible forms of entertainment.

Table 4.1 Televisions in use

	Televisions in use (thousands)		
	1977	1988	% increase
Argentina	4 600	6 850	48.9
Uruguay	360	535	48.6
Mexico	5 480	10 500	91.6
Brazil	11 000	28 000	154.5
Chile	710	2 330	228.2
Ecuador	340	825	142.6
Peru	825	1 800	118.2
El Salvador	148	425	116.7
Guatemala	150	325	91.6

Source: Derived from *The International Marketing and Data Statistics*, 16th edn, 1990.

The debate over the role of the media is similar to the master–servant debate over bureaucracy. Is the media merely a tool that can be used by the dominant group in society to bring about whatever change or lack of change they require or does it, by its nature, shape the influence it has on the population? The discussion is closely linked to the different approaches of modernisation and dependency theories.

Table 4.2 Radios in use

	Radios in use (thousands)		
	1977	1988	% increase
Argentina	10 000	21 000	110
Uruguay	1 625	1 835	12.9
Mexico (1980)	9 000	20 500	228
Brazil	16 980	53 500	215.1
Chile	2 000	4 308	115.4
Honduras	163	1 847	1033.1
Peru	22 000	51 300	133.2
El Salvador	1 415	2 040	44.2
Guatemala	275	550	100.0

Source: Derived from *The International Marketing and Data Statistics*, 16th edn, 1990.

The modernisation school sees the media as, on the one hand, an aspect of modern society and, on the other, a means by which élites can modernise further. Joseph Kahl, in his attempt to measure modernism, picks out participation in the mass media as one of his key criteria of the modern. For him, modern 'man' reads newspapers, listens to the radio, watches television and follows national and international events. In his research on Mexico and Brazil, he found that a good proportion of his samples were interested in following the national and international news both on the radio and in the newspapers

The power of the media as a mechanism for change lies in the scope and breadth of the population it reaches. Often referred to as a means by which a nation, or even the world, can become a community, it makes possible the transmitting of messages from a small group to a vast number of people. Television and radio can be used for educational purposes to spread modern values and to inform people. In Mexico and El Salvador television has been used experimentally as an educational medium and, in Brazil, classic novels have been serialised to give people a knowledge of their own culture. In theory, the media can be used to promote certain policies though, in practice, this is not often the case, partly because broadcasting, the obvious channel, needs to attract advertising to finance itself and, since educational programmes do not often have mass appeal, they are unlikely to draw the necessary funding. The Cubans, however, with the help of state resources, have used television and film to encourage the development of the values of the revolution, especially where these are at variance with traditional and deeply held views. Television has been used to promote equality between the sexes by showing short features suggesting that sons and daughters should be brought up equally at home and girls not expected to wait on their brothers.

Taking the perspective of the dependency school, mass media in the LDCs is an aspect of cultural imperialism. The technology necessary to produce films and television and radio programmes, is so much more advanced in industrialised countries than in the LDCs that it is frequently much simpler and cheaper to import material than to produce it in the Third World. Although countries like Chile, Mexico and Brazil have budding film industries, the products of which are now reaching the industrialised world, it is still North American and European films that predominate in the cinemas of Mexico City, Santiago and Sao Paulo. Brazilian cinema has expanded rapidly, for its share of the domestic market has risen from 14 per cent in 1970 to 38 per cent in 1983 (*LAWR*, Brazil, November 1983); nevertheless, this still leaves nearly two-thirds of the market to films from abroad. Television and radio in the region have several channels so, while some are devoted to local issues and productions, others are packed with imported serials, films and games. Not only are foreign programmes often better made, but they also have the allure, real or imagined, of being sophisticated or trendy. The point about imported material is that it contains the values of the society where it was made and, so, through modern techniques of diffusion, all social levels of a Third World's society are exposed to the product of a small group of people in the industrialised world. The world becomes a community, which is influenced by the small élite who control the media. 'In a dependent society the media demonstrate the imperialist system's concept of change – a conception which in fact ends up being the denial of change' (Mattelart 1978: 23).

The implicit values in drama are made more explicit in the advertising that sponsors and accompanies the fiction. In promoting consumer products, in competition with each other, the advertisers promote the values of capitalism – the stress laid on wealth and working hard in conventional modes to try and achieve it, competition and the importance of consumer goods as an expression of self and status. According to Mattelart, North American advertising agencies in Latin America have gone one step further by explicitly promoting models of political development (Mattelart 1978).

Although it is clear that the potential for considerable influence is there, how effective is the mass media in changing or maintaining values and behaviour patterns? A study of drinking patterns in Mexico reveals the strength of advertising. Mexico has one of the highest rates of alcoholism in the world, which appears to be the result of the coincidence of traditional and modern advertisement-inspired modes of behaviour. Mexicans have always been known for heavy drinking because traditional forms of alcohol are cheap, peasants often making their own, and this is the classic way of forgetting the drudgery of poverty. But in recent years alcoholism has become a greater and nationally recognised social problem. The figures show that beer and spirits together constitute the industry with the greatest publicity in Mexico (Harvey 1983). Those with a little cash to spend, who want to be modern and sophisticated as the advertisements suggest, buy imported whisky and brandy. As a result, alcoholism affects all levels of society and all age groups.

Not only visual representation, but the written word, too, is subject to the influence of imperialism. Latin American newspapers, unable to afford their own reporters in different parts of the world, often use syndicated news from North America or other foreign sources. This bears the imperialist nation's view of world events and is then reproduced for the mass of Latin Americans to read. As well as newspapers, and for those whose literacy level is not up to the language of the newspaper, comics for adults are widely available. An aspect of popular culture that is not found much in the industrialised world, comics are read by all levels of society and all age groups.

CONCLUSION

The traditional Catholic Church has lost its long-term dominance of the value system, giving way to the spread of new ideas from the secular world and indeed to new perspectives and views from within its own ranks and from other religions. In general the changes taking place within the institutions which influence values reflect a gradual shift towards the norms of contemporary capitalist society. The ideas that the media purveys are certainly both modern and an aspect of cultural imperialism, for they emanate from the core of the countries of the capitalist world. Education has expanded very significantly, thus giving a greater number of people information and skills to assist them in improving their living standards. Health care has grown as the result of the spread of values that associate health care with development, and provision of welfare services with a modern democratic state.

The impact of changes in values has been limited by a lack of corresponding structural change. This suggests that economic and political factors are vital in shaping the nature and extent of the effects of changes within these institutions (an argument that will be examined more closely in Chapters 8 and 9).

FURTHER READING

Albornoz, O. (1993) *Education and society in Latin America*. Macmillan, Basingstoke and London.

CEPAL Review, 21, December 1983.

Archer, D., Costello, P. (1990) *Literacy and power: the Latin American battleground*. Earthscan publications, London.

Green, D. (1991) *Faces of Latin America*. LAB, London.

King, J. (1990) *Magical reels, a history of cinema in Latin America*, Verso London.

Levine, D. H. (1981) *Religion and politics in Latin America: the Catholic church in Venezuela and Colombia*. Princeton University Press, Princeton Guilford, Surrey.

Martin, D. (1990) *Tongues of Fire, The explosion of protestantism in Latin America*. Basil Blackwell, Oxford.

Navarro, V. (1976) *Medicine under capitalism*. Croom Helm. London, New York.

Smith, B. H. (1982) *The church and politics in Chile. Challenges to modern catholicism*. Princeton University Press, Guilford, Surrey.

Chapter

5

Social relations

Despite the spread of values emphasising achievement and competition, personalism remains an essential feature of social relations in Latin America, probably because of the continuance of uncertainty and insecurity. Life in an Indian community, over the generations, is characterised by a lack of control over the essentials. The subsistence farmer relies on his produce to feed his family, but he has no control over the weather and the conditions of the soil, which are so critical, he does not have recourse to fertilisers, pesticides and equipment to counteract the malevolent side of nature. The reason for illness and death are unknown to him so he can do nothing to alleviate the high rates of infant mortality. A genuine lack of control contributes to feelings of fatalism and passivity: after all, what can he do to improve the situation? He is dependent on fate. If hunger prevails, all he can do is kill a few chickens which, with luck, he has kept for such an occasion, to ease the situation temporarily, but he has no long-term solution at his fingertips. Long-term planning is, therefore, alien to the Indians living in these conditions. Moreover, since accident and misfortune are not fully understood, this is often attributed to witchcraft or magic, brought about by hostile feelings of others. In such a situation it is important not to make enemies, so peasants are reluctant to become too involved with others and are particularly suspicious and wary of newcomers. General insecurity, the inability to direct one's life in a positive way and a general fear of the unknown, and therefore strangers, all contrive to produce a disposition that does not favour the activities of planners and developers. In this way, the attitudes of Indians in isolated communities are very similar to those of peasants living in the same sorts of conditions throughout the world (Bailey 1971).

In Latin America this insecurity and the uneven distribution of scarce resources leads to an exaggerated concern to establish warm and useful relationships with others. For the mass of the people jobs are badly paid and irregular, good quality land is in short supply, and the mechanism of power in society is totally beyond their comprehension. Thus few can be sure of the resources they need to support themselves and their families.

Even the fortunes of the rich can oscillate significantly, due to unstable political and economic situations and changing local administrations. In such an environment of uncertainty, people need something they can rely on and they turn to people they can trust. In a situation where inflation can rocket to over 100 per cent, investing in personal relationships, by establishing good friends who can help in a crisis, may be more useful than putting all one's savings into cash. For the peasant, whose crop may be hit by blight or drought at any time, the only source of help and support often lies in those close to him. 'In Latin America, people invest much of their effort in consolidating their personal fields. They increase their security in a narrow orbit to counteract insecurity in the wider orbit' (Wolf and Hansen 1972: 200).

Explanations of fatalistic and passive attitudes in terms of a culture of repression (Huizer 1973) and image of limited good (Foster 1967) are rarely referred to now as many sociologists and anthropologists have come to challenge the view that the peasantry is conservative and reluctant to change. Peasants in the 1990s have shown that they are quite willing to experiment with what for them are new forms of technology. Huizer points out that the distrust of the peasantry can be overcome if a genuine attempt is made to understand the peasants' situation and to explain fully the aims and methods of the development project. As part of the project sponsored by the UN World Food Programme and the Chilean government to develop the agricultural potential of sixteen Chilean communities, Huizer and the rest of the team encountered initial resistance on the part of the local peasantry, who had in the past lost land to local land-owners in the area and as a result, were reluctant to trust strangers. Huizer encouraged the peasants to discuss their problems, and having learned that their main grievance was the loss of land, he won their confidence by supporting them in this issue against vested interests. Then management of the World Food Programme facilities at local level was transferred to the elected representatives of the peasant community, thus involving them positively in the project. An attempt to see the situation from the point of view of the peasantry and to deliberately alleviate their fears had turned hostility into participation. The consequence was that in the following few months 'seven new roads were improved or repaired, eight schools were built, five irrigation canals were constructed or repaired, 2000 lemon trees were planted and steps towards the

formation of multi-purpose peasant co-operatives were made' (Huizer 1973: 28).
Various Indian communities have demonstrated a flair for entrepreneurship from within their own ranks, which shows that conservatism is by no means universal. The Ecuadorian Otavalo, renowned for their weaving, sell their ponchos not only in other Latin American countries, but even as far afield as Europe.

PERSONALISM

Relationships between two people anywhere in the world can be expressed in terms of a dyadic contract (Foster 1967). The formal institutions in society provide the individual with a framework and, within this, he/she can choose with whom to interact, thus setting up a dyadic contract, i.e., an informal relationship that binds two people. Although informal, this relationship can be recognised through reciprocity. Whether it be through the exchange of goods, services, esteem or confidence, all relationships are marked by some sort of reciprocity. The exchange of food and drink through dinner parties, coffee meetings, pub evenings, etc., is a universal means of establishing and maintaining social relations. In Latin America relationships are built up in this way not just for personal support but also to promote commercial and political ends.

The Brazilian *panelinhas* are informal groups made up of a number of dyadic contracts, i.e. they are people linked by personal ties – family, kin, friends – but the members are selected according to their occupation. Each *panelinha* tries to represent a range of occupations, such as doctor, lawyer, banker, engineer, architect and a variety of different businesses. Members of one group will try and do business with each other, rather than outsiders, knowing they have the advantage of the personal contact and can, therefore, trust the person concerned (Leeds 1974). It is not just in professional and business circles that personal contacts help people obtain jobs. From her Mexican study, Eckstein reports that about 38 per cent of the men with factory and white-collar jobs said they obtained their work through a personal acquaintance (Eckstein 1977).

Dyadic contracts can be divided into colleague contracts, referring to relationships between those of equal status, and patron–client contracts, where the status of those involved is different. Colleague contracts, such as most friendships, involve an exchange of similar kinds of things whereas, in the patron–client relationship, the different status is reflected in the fact that the patron offers something quite different to the relationship from the contribution of the client. One of the most widespread expressions of the colleague contract is *compadrazgo*.

Compadrazgo, as Latin Americans call ritual or pseudo-kinship, is a means by which people imbue relationships of friendship with the lasting and obligatory character of kinship. Friendship has the advantage of choice – friends are chosen by the individual concerned – but it is also flexible and not necessarily lasting. Kinship is binding and permanent but permits no choice of personnel; the individual must accept the relatives he has. *Compadrazgo* is a compromise which attempts to combine the best of friendship with the advantages of kinship (Foster 1967).

Compadrazgo relations are set up through the ritual sponsorship of the church system by appointing godparents for children. On at least four ritual occasions in the life of a young person – baptism, confirmation, first communion and marriage – the parents have the opportunity of asking friends to sponsor them, thus becoming their godparents. This establishes a set of permanent relationships between parents, godparents, and child/godchild in which the most important social relationship excludes the child, for it is between parents and godparents. They become *compadres*, or ritual kin. The godparents may have a ritual role to the child, such as participating in his wedding ceremony, and taking on certain responsibilities regarding his religious education, but the relationship that is strongest and requires respect, warmth, and the obligation to help at all times is the one between parents and godparents.

In the task of fortifying oneself with supportive personal relationships, *compadrazgo* offers extensive manipulative opportunities. Each ceremony for each child requires a new set of godparents so, as most Latin American couples have many children, there are numerous opportunities for establishing new sets of *compadres*. Moreover, one can extend secondary relationships to the kin and *compadres* of a *compadre* with whom one has a primary relationship. In the early years of married life, a couple may try to establish as a wide social network as possible and, then, in later life if they feel that it is sufficient, they may ask relatives to be godparents, thus reaffirming existing kinship ties rather than extending the network further. At the level of village life, this pattern provides stability and cohesion, for villagers are interlinked by the criss-crossing relationships of *compadrazgo*.

In urban areas *compadrazgo* is more informal though in some areas just as effective. There is, however, considerable regional variation. In Mexico, especially in rural areas, it operates as an important principle of social organisation, while in Chile it has lost a lot of its formal and obligatory nature and is not very meaningful for many people.

Clientelism, however, continues to be a firmly entrenched pattern of social relations throughout the region. The exchange of dissimilar things in the patron–client relationship occurs because the patron offers the qualities that accrue to his superior status, that is, power and influence to help the client, while in return the latter contributes support and esteem. From this the client gains assistance which he needs in his weak position. The peasant, for instance, who wishes to sell his avocados or pots in the local market may need a licence to do so, something that is outside the realm of the usual activities of a peasant, especially an illiterate. His. patron, usually the local landowner, can acquire the necessary application forms and show him what to do with them. For his part, the patron gains the personal support of his client and, when he has accumulated a number of these clients, he becomes a prestigious person in the area due to the cumulative honour from them. This support can be turned into power by an astute patron. The patron can call on his clients for active support, a facet which is important in local power struggles. Candidates in local elections can expect their followers not only to vote for them but to campaign for them as well. Political power can be built up in this way, with the result that many Latin American leaders owe their position to a network of clientelist relationships.

Although the status difference may be emphasised in the style of the relationship, with client addressing the patron as *usted*, but the latter using the more familiar *tu* the whole relationship is imbued with a personal element. Patron and client may be *compadres*, in which case the patron is expected to take an interest in the child. The initiative to set up the relationship often comes from the client, though at times this is difficult because he has to approach someone whose social position is very different to his own. In this case, he may resort to the use of a *palanca*. Literally meaning leaver, the *palanca* is the client's means of access to the patron, an intermediary between the two. While studying Tzintzuntzan, the anthropologist George Foster found that the female villagers would ask his wife to approach him for a favour (Foster 1967).

It may seem so far that in terms of clearly defined benefits the client comes out best from the deal. However, in the long term this is not the case, because of the extent of power that can be gained by patrons. The client gets help for day-to-day problems, but the price he pays for this is to be the landlord's follower, to do his bidding and accept his view of the world. In this way a small number of landowners can control the peasantry in rural areas. It is a way in which the dominant class of landowners manipulates a subordinate one. Through clientelism, the allegiance of the peasant is to his patron rather than to the other peasants, so that it is more difficult for solidarity between those in the peasant class to develop. The structuring of loyalties vertically in this way inhibits the development of horizontal bonds and thus class identity among the poor. Moreover, clientelism prevents the peasant making contact with the wider society. As the landowner acts as an intermediary, the peasant gains no experience of wider society and must depend on his patron for his interpretation of the world. The essence of clientelism requires class differences; thus, although it softens the harshness of status differences it also ensures their continuation.

Far from declining with industrialisation and urbanisation, clientelism has flourished. Many businesses, including large and successful ones, are based on the personal relations of clientelism. SIAM is a large Argentine industrial-commercial complex which was run by Torcuato Di Tella, as a family business. Those at the top level had Di Tella's confidence through a relationship of kinship or friendship, and links between middle and upper management levels were based on personal relationships. Di Tella took a personal interest in the lives of his workers, giving presents at weddings and on the birth of children and taking responsibility for the health and welfare of the workers and their families. In return, he could count on their loyalty and support for him and the firm. There was little delegation of authority, but Di Tella's personal participation enhanced his authority and gave him considerable control. This style of management continued after his death (Wolf and Hansen 1972).

Rothstein argues that patron–client relationships are well suited to the peripheral capitalist development that has taken place in Latin America because they are a way of paring down the number of recipients of industrial gains when these gains are few in number. Economic growth in Latin America has accumulated benefits in the hands of a few, rather than distributing them, as predicted by the modernists. The small surplus that is distributed is so limited that it cannot be shared by all, but patronage is a mechanism of selecting a few workers to benefit. Clientelism is a strategy used by capitalists and workers to adapt to a situation where there is limited mobility. For the worker, patronage yields access to some of the benefits of industrialisation and, for the manager, it is a way of distributing a small number of benefits, at the same time assuring support and a disciplined labour force (Rothstein 1979).

FAMILY AND HOUSEHOLD

The most obvious place to turn to for close ties is the family. The family offers the potential for a large number of links and is a source of permanent relationships, even if conditions around one are changing. Large families and the recognition of a wide network of kinship ties is a feature of Latin American life. Expansion of this network is greatest among the rich, where people may retain contact with a hundred or more relatives. In this connection, family names are important. Latin Americans maintain the Spanish custom of adding the mother's surname after the father's, thus denoting both families of origin. In some countries, names demonstrate aristocratic status and the élite may be referred to, as it is in Peru, as the forty families. But the family among the rich is not only a source of prestige: it also has a very functional aspect, since family networks are manipulated in the course of politics and business.

Some of the most modern industrial companies in Latin America are family businesses, whose boards of directors are made up of relatives and close friends who can be trusted. Businesses are sometimes seen as sources of employment for family members. The owner of a large concern may well feel an obligation to find a position in his firm for his hard-up cousin. Above all, the family is a property-owning unit, whether the property is a business, estates, land or houses, so that these resources are controlled and acquired by succeeding generations through the family.

For the poor, the family is not a mechanism to control resources, but an institution to turn to because of the scarcity of resources. The scope of their kinship networks is smaller but just as significant because they provide help in times of hardship. Peasants migrating to the city usually turn to relatives for help in the new and frightening environment. It is kin, often, who provide shelter during the first days in the town and who help the migrants build their own homes. It is they, too, who lend them money for the journey to help them look for jobs in the urban environment. The migration process frequently involves a chain of relatives from rural village to urban neighbourhood.

Traditionally the family was patriarchal and authoritarian. The head of the family frequently had under his control a large number of people, both relatives and servants, and considerable resources. The best documented examples come from Brazil, where the patriarchal family appeared after Portuguese colonisation with the agrarian families. The patriarch's power over his family and offspring was absolute and often rigorously exercised. The strength of the institution can be seen in the fact that it lasted for generations in the sugar estates, despite social changes taking place around it.

The strongly authoritarian role of the head of the family has been undermined by the opportunities created by urbanisation for young people to be independent and earn a living outside the family, and by conceptions of the North American family as portrayed by the media. Nevertheless, the family remains patriarchal and a key institution in society. The proposition of North American sociologists such as William Goode that urbanisation leads to the decline of the family is not fully supported in the Latin American context (Goode 1963). The original argument suggests that the family declines both in terms of size and function. The extended family gives way to the nuclear and in so doing loses many of its crucial functions. The proposition is that urban life requires a flexible and therefore preferably small domestic unit because of the need for mobility. The family needs to be geographically mobile to move to places of work and therefore the fewer the members to uproot the better. Social mobility is also made easier when there are fewer members associated with previous social positions. Moreover, a major characteristic of urban living until very recently has been the way the family is usually supported by one breadwinner. In this context, larger numbers mean more mouths to feed without the corresponding contribution to family subsistence, as happens in the traditional rural family. In terms of function, the family in the countryside is both a consumption unit and production unit. The latter declines in the urban setting, leaving the family with only its consumption role (Goode 1963).

Material from Latin America both supports and contradicts Goode's view. Although the nuclear family predominates as the main household unit, there is considerable evidence to show the importance of the extended family, and although the family production unit may be on the wane, other economic functions have taken its place. Frequently, too, the extended family remains an important supportive unit.

Chant's study of a low-income district of Querétaro, Mexico, suggests that the extended family may be more suitable than the nuclear for urban living (Chant 1991). Querétaro relies on heavy industry, with few opportunities for women, but many women do operate in the informal sector. Chant found that the extended family, whether male- or female-headed, was better able to support women taking on remunerative work and therefore had higher earnings than the nuclear families in her sample. This was because the members of the extended family could support each other over child-care and housework, allowing respective members to work outside the home. The higher-earning families were able to gain local prestige by spending on their homes. The nuclear family, however, is conducive to a strict segregation between male and female roles, which inhibits women from entering the labour force. The fact that the extended families were the high earners demonstrates a very real economic function.

Rather different evidence from Chile shows the increase of the extended family in urban areas. Both the economic crises of the 1980s and government policies have led to a housing shortage in Santiago, with the result that young people are unable to set up their own households. The average number of persons per household in 1986 was 6.6, compared with 5.6 in 1966 (Martinez and Valenzuela 1986). The researchers found that over 50 per cent of the families interviewed were extended, usually married children living in their parents' homes.

Rural–urban migration in Peru has in some areas helped to reinforce the extended family through the setting up of confederations of households (Alderson-Smith 1984). Migration to Huancayo and Lima from the Montaro valley involves linkages between households, which the author calls 'federations'. These linkages are supportive, in that migrants usually stay first with kin in the town, and also economically beneficial since confederations are a means of minimising risk and uncertainty by embracing different forms of productive enterprise over a wide area. Federations fragment and reformulate in relation to changing economic requirements of each household. In many cases, brothers with separate families own livestock together, in others strawberries are grown in the rural areas and family members move to kin in the town for the strawberry-selling season.

The family is a flexible and adaptable institution which changes to meet the needs of the new urban situation. In many cases the larger extended unit is the most appropriate form, and the new functions, though specifically different are invariably economic and supportive.

The importance of economic functions for the family can be seen in the attention recently focused on 'household survival strategies'. Despite increasing reference to this term, the definition remains problematic and varies between authors. The common elements, however, include a group of related and/or unrelated individuals who live under the same roof and constitute an economic decision-making unit. It is a social organisation that organises resources and recruits and allocates labour in a combination of productive and reproductive tasks. Households can choose between the production of use values, the production of commodities and the sale of labour for a wage. The objectives of any household are to ensure its own reproduction as a social unit and to improve its material welfare, both of which entail the development of strategies.

Chant points out that one of the main advantages of the extended family is the opportunities it gives women for participation in the labour force, with the implication that labour force participation of women considerably augments household income. Female labour force participation is therefore a key aspect of survival strategies and has proved positive in combating poverty among low-income groups (Chant 1991). Economic crisis and structural adjustment have also encouraged the premature entry of sons into the labour market, with daughters frequently taking on domestic labour at an earlier age at home. It is significant that participation in the labour force by women rarely results in male partners sharing domestic tasks.

How households use available resources – the time, skills and energy of their members – depends on how they perceive their environment, the goals to which they aspire, and their expectations about the benefits they will receive for their effort. Household strategies will depend on the number of earners, age and sex of family members, who controls household budgets, and how household earnings are allocated.

The household-strategies approach has been useful in analysing migration patterns. It has drawn attention to the significance of tasks associated with household maintenance, so often overlooked with the focus on wage-labour opportunities, in explaining gender-differentiated migration from rural households. Gender divisions of labour form a basic fabric for the household which provide the basic underpinnings for decisions about who should migrate and who should stay.

Much of the family's strength can be attributed to vigorous support from the church, which has always placed the family at the centre of Christian living. The church's view, that marriages sanctioned by religious ritual should not be broken, is reflected in divorce laws. In several Latin American countries there is no legal concept of divorce. In Mexico, however, where the state broke with the church quite violently during the revolution, divorce is possible (and, indeed, in some places very easy). The church's opposition to mechanical and chemical contraception and to abortion greatly encourages the large families that are so characteristic of the region.

The population explosion has been one of the major phenomena of Latin America this century. Since 1950 the population has been steadily increasing from 164 million to 422 million in mid 1991. This is very much a Third World phenomenon for, throughout these areas, the rate of population increase is considerably higher than in the rest of the industrialised world. For the period from 1985–90, the annual rate of population increase was 2.99 per cent for Africa, 1.87 for Asia, 1.48 for Oceania and 2.01 for South America. Europe, however, during this period had an average annual rate of only 0.25 per cent (World Resources 1993: 246). Even for LDCs, the Latin American rate is high.

Although, as suggested in Chapter 1, rapid population increase is not a major development problem, it is a cause for concern in the region. It has meant that in Mexico, for instance, although the proportion of illiterates nationally has decreased every decade, in recent years absolute numbers have been increasing. Birth control, however, is a very controversial issue. Some programmes set up by North American and international development agencies in order to alleviate the population pressure on food and thus reduce poverty were aimed at direct population control rather than giving women the right and the freedom to control their own fertility. These have gained a reputation for testing unsafe contraceptives on poor Latin American women and enforcing sterilisation. The Peace Corps, the US volunteer programme, has been expelled from both Bolivia and Peru for its activities in sterilising peasant women without their knowledge (Bronstein 1982).

Fear, partly due to rumours about these sterilisation programmes, and ignorance prevent the majority of Latin American women from using contraceptives. The attitude of the church has made it difficult for governments openly to promote birth-control programmes that do anything more than advocate natural methods. In Mexico the government has encouraged the spread of contraceptive methods through television advertisements advocating the advantages of small families; but it has been forced, because of the influence of the church, to take a low profile and operate very discreetly.

Throughout Latin America, except for Cuba, abortion is illegal; nevertheless, a considerable number of women do resort to abortion, which is often carried out in inadequate and unhygienic conditions.

The World Health Organisation (in reference to Brazil) has estimated that for three to four million women per annum the only solution to an unwanted pregnancy is an illegal abortion carried out in the majority of cases under dangerous conditions. Poorer women more often carry out their own abortions with the help of poisonous herbs, by injecting themselves with several doses of hormonal contraceptives (available in almost every chemist's shop), by inserting pointed objects or probes, or injecting acidic liquids into the womb. As a consequence, 200,000 women each year are taken into hospital after abortions with serious sometimes fatal complications.

(Caipora Women's Group (CWG) 1993: 83)

For Latin America as a whole, Bronstein (1982) suggests that 30–50 per cent of all maternal deaths are from improperly performed abortions.

GENDER AND DEVELOPMENT

Undoubtedly rapid population expansion in Latin America can be traced in part, to *machismo*. Taken from the Spanish word *macho*, meaning male, *machismo* refers to an ideology of behaviour in which distinctly different lifestyles are deemed proper for men and women. *Machismo* is an exaggerated cult of virility which expresses itself in male assertations of superiority over females, and competition between men. To fulfil *macho* behaviour patterns a man must show no fear, demonstrate sexual prowess, father many children and exercise tight control over any female kin. Adultery is seen as natural for a man, but a serious offence for women. *Machismo* is a New World phenomenon with roots in old world cultures. Concepts of honour and shame associated with manliness can be found in many of the cultures of southern Europe, but while this behaviour seems to have declined in Spain it has flourished in Latin America. Since its early expression was in the form of sexual relations between conquering white soldiers and dominated Indian women, it has taken on a distinctly aggressive element which can emerge in the form of violence. It is a characteristic of *mestizo* Latin America, for studies of Indian communities often reveal a relative absence of *machismo*. Its counterpart is *marianismo*, the submissive female role. The ideal woman is gentle, kind, long-suffering, loving and submits to the demands of men, whether they be husbands, fathers, sons or brothers. She has an infinite capacity for humility and sacrifice and an abundant store of patience. In one area, however, she does come into her own, since she has spiritual strength and moral superiority. This strength makes her the centre of the family and the one responsible for keeping its members together. The role model for the woman is the Virgin Mary, hence the term *marianismo*.

A major concern today is the participation of women in development. Previously it was argued that women were left out of development, as revealed by the dates of women's suffrage in the region (See Table 5.1). Today women participate fully in the process of development but do not receive benefits commensurate with their activities. 'The problem lies in the imbalance between their contribution and the benefits they receive, an imbalance which is even greater if their potential contribution is compared with their actual benefits' (López and Pollack 1989: 37).

Greater participation by women has been due to a number of factors. In recent years educational levels of women have risen, making them

111

better qualified to participate in the labour market. In fact, in Brazil, Colombia, Nicaragua and Panama the percentage of female students enrolled in tertiary education is higher than that in the United Kingdom (*The Economist Book of Vital World Statistics*, 1990: 209). In Costa Rica educational achievements are better among women than men, although men remain the higher earners. The percentage of female literates still remains less than that of men in most Latin American countries; however, the number of female literates has shown marked improvements over the past decade (see Table 5.2).

Table 5.1 Dates of women's suffrage in Latin America

Ecuador	1929	Chile	1949
Brazil	1932	Costa Rica	1949
Uruguay	1932	Haiti	1950
Cuba	1934	Bolivia	1952
El Salvador	1939	Mexico	1953
Dominican Republic	1942	Honduras	1955
Guatemala	1945	Nicaragua	1955
Panama	1945	Peru	1955
Argentina	1947	Colombia	1957
Venezuela	1947	Paraguay	1961

Source: Nash and Safa, *Sex and class in Latin America*. 1976: 223.

The resulting increased participation in the labour force, which will be dealt with later in the chapter, has in itself led to development activities. The extensive urbanisation in the region has meant a migration of women to cities which is many times higher than that of men.

The trend for a greater percentage of households to be headed by women as a result of migration and new family organisation patterns has thrust women into a participatory role as both breadwinners and defenders of private family interests. At the same time it reveals a 'feminisation of poverty' as the percentage of households headed by women is much greater at the indigent levels than at others (see Table 5.3).

The fact that the majority of female-headed households at indigent levels are also women of Indigenous or African descent must not be ignored as a factor of analysis. 'Latin American feminism loses much of its force by making abstract of a fact of great importance: the pluricultural and multiracial character of societies in the region' (Gonzales, cited Radcliffe and Westwood 1993: 5). The oppression of women from Latin America is a generality which disguises the hard reality lived by millions of Black and Indigenous women. In Brazil, for example, 37 per cent of Black women are the primary source of their families' incomes, compared to 12 per cent among white women. Their average monthly income is about US $50, only one-third of the average income of households headed by white women (Caipora Women's Group 1993: 12).

Table 5.2 Adult male and female literacy in Latin America: 1970 and 1990

Country	Adult female literacy (percentage)		Adult male literacy (percentage)	
	1970	1990	1970	1990
Costa Rica	87	93	98	93
Cuba	87	93	86	95
Dominican Republic	65	82	69	85
El Salvador	53	70	61	76
Guatemala	37	47	51	63
Haiti	17	47	26	59
Honduras	50	71	55	76
Mexico	69	85	78	90
Nicaragua	57	×	59	×
Panama	81	88	81	88
Argentina	92	95	94	96
Bolivia	46	71	68	85
Brazil	63	80	69	83
Chile	88	93	90	94
Colombia	76	86	79	88
Ecuador	68	84	75	88
Paraguay	75	88	85	92
Peru	60	79	81	92
Uruguay	93	96	93	97
Venezuela	71	90	79	87

Source: Derived from World Resources 1993: 254.

Table 5.3 Heads of household by level and sex

		Households		
	Total	*Indigent*	*Poor*	*Non-poor*
Costa Rica				
Men	84.3	62.7	82.3	86.2
Women	15.7	37.1	17.7	13.8
Venezuela				
Men	88.3	63.7	86.4	89.4
Women	11.7	36.3	13.6	10.6
Chile				
Men	89.3	87.6	90.4	89.6
Women	10.7	12.4	9.6	10.4
Peru				
Men	88.5	76.7	90.3	91.3
Women	11.5	23.3	9.7	8.7

Source: López and Pollack, *CEPAL Review* **39**, December 1989: 43.

The need to press for greater benefits has become apparent to women who have begun organising in a number of different ways: a process which has led them out of the private sphere into the public. In Latin America cultural definitions of social space have always been deeply gendered. Domestic space is primarily associated with women's activities and the public domain with the male world. Clearly these categories are not exclusive and there has been much debate about the value of the private/public distinction. It has an immediate common-sense application, but needs to be refined to be a useful tool of analysis. In general terms, however, it can be seen that some of the issues normally associated with private space have drawn women into the public arena.

Conditions in the rapidly expanding cities have produced difficult situations which often have a greater immediate impact on women than men. Poor urban facilities, such as water shortages, mean that women do not have water for their children to drink and wash. Rising food prices are initially felt by women trying to feed families on low incomes. Under such circumstances, women have a clear reason to take action, their activism being rooted in their desire for their families' well-being. Logan argues that women often have more flexible work schedules than men, which allows them to participate in community activities (Logan 1988). In many places community involvement has mushroomed into protest movements, so leading women into confrontation with local authorities as they attempt to negotiate and argue for the required improvements. Political activism has always been associated with public space and the male world. It is worth noting that most of the research shows that these urban movements, with one or two notable exceptions such as the *madres* in Argentina, have male leaders, though women may well form the bulk of the membership.

Nowhere has this move into the political sphere in the defence of family interests been so clearly seen as in the southern cone countries in the 1970s. Here motherhood, which often operates as a controlling mechanism keeping women in the narrower private world, becomes a politicising mechanism. All that need be said to demonstrate the controlling aspect is summed up in the words of General Pinochet, head of the military junta and President of Chile (1973–90): 'Woman, from the moment she becomes a mother, expects nothing more in terms of material things: she seeks and finds the purpose of her life in her child, her only treasure, and the object of all her dreams' (Pinochet, *El Mecurio*, 19 October 1979).

'It was precisely in the name of defending the family and children that women began taking on a more public role' (Fisher 1993: 179). Every Thursday afternoon a group of women wearing white headscarves gather in the Plaza de Mayo in the centre of Buenos Aires to walk in a circle around the monument to independence. They are protesting at the disappearance of an estimated 30 000 young children and babies at the hands of state-organised death squads responsible for kidnappings, torture and murder (Fisher 1993). Although the generals responsible have now fallen from power, the protest continues, aimed at some just punishment for the guilty. The mothers' vigil started in 1977 when fourteen women gathered in the square to demand information about their missing children. Within months they were joined by dozens of other women, becoming the first group to challenge publicly one of the region's most brutal military regimes and eventually a key element in its downfall.

The role of motherhood, usually a pillar of the private world, is in this case the vehicle of transition into the public arena of national politics. The reason for protest is a family one and the method of protest, non-violent quiet and passive, also associated with the spiritual qualities of motherhood. The sanctity surrounding the role offers some initial protection in the public arena, though it has to be pointed out that in Argentina, as in other Latin American countries, that protection is very limited. Mothers here, as in Chile and Central America, have been abused and murdered by security forces. Motherhood, however, became a very effective and powerful political weapon. The *madres* of the Plaza de Mayo achieved international fame which helped to undermine the standing of the military junta. They brought to the attention of Argentines the brutalities of a government which professed to stand for family values.

The kinds of actions demonstrated by women involving themselves in community action for better facilities and by the southern cone *madres* reflects a growing consciousness among women. In analysing these developments a refinement of the public/private distinction has been made by Maxine Molyneux using the concepts 'practical gender interests' and 'strategic gender interests'. The former refers to interests which arise from women's positions in sexual divisions of labour such as interests as mothers or rural labourers. Strategic interests arise from a desire to challenge women's subordination and existing gender relations (Molyneux 1985). The awareness of strategic interests has led to some women's movements particularly devoted to changing women's positions in society, such as the women's section of National Council of Urban Popular Movements (CONAMUP) in Mexico, but they remain limited in number.

Moser has adapted Molyneux's concepts to the area of needs. She uses practical gender needs to include sanitation, children's nurseries, higher wages, health care, programmes for women and other similar items. Strategic gender needs might include issues such as changes in divorce laws to give women equality with men, or 'affirmative action' to give women political representation (Moser 1989). They might also include laws that recognise the effect that racial inequality has on women. This allows us to use the practical/strategic distinction to refer to problems and requirements.

If women are to benefit more from their contribution to the development process they need to move towards a greater awareness of strategic needs and interests. They are participating more in the public arena, though mostly in defence of practical interests. This takes them some of the way because it means women taking on new roles and identities and developing an awareness of the need for more fundamental changes in power structures and in relations between men and women. The crucial step occurs through a process of empowerment. Empowerment involves moving from individual to collective action, increasing capabilities and challenging existing structures. Some of these elements have been achieved, though there is still a long way to go before full empowerment is reached (Johnson 1992).

Any question of development involves a discussion of participation in the labour force. Prior to addressing this issue it is important to point out that statistics representing participation rates of women in the workforce are notoriously inaccurate. Women are often statistically 'invisible' because they work in areas that are not recorded as full-time occupations. Much of womens's rural work, such as caring for poultry, milking and craftwork within the home, is not included, since men are registered as the family breadwinners, as farmers, and women as housewives. In the same way, 'home industries' of women in the towns do not find their way into occupational statistics.

A post-census study in Costa Rica reveals the weakness of formal figures. In 1983 all the women in the district of San Juan in San Ramon de Alajuela aged twelve or over who had been classified by the census that year as economically inactive were reinterviewed. The researchers found that 25 per cent of the 'inactive' women had worked for pay during the reference week and almost 60 per cent had worked seasonally during the coffee harvest. More than 40 per cent of the 'inactive' reported working six months of the year or more (Buvnic and Horenstein 1986). The tendency for many women to consider themselves primarily housewives and record this in a census obscures the extent of paid work carried out by women. Official statistics, therefore, can only be used as indicators, rather than accurate statements, but they do offer some guides to the occupational picture.

Equally, due to an internalised conception of racial social status, people will often identify with groups a few stages whiter than themselves when records of ethnic composition of the populations are being made.

GENDER AND THE LABOUR MARKET

Changes in structural factors have taken place in the form of peripheral capitalist industrialisation, which has had considerable impact on the participation of women of all classes and races in the labour market. Extent of labour-force participation is an important indicator of the position of women because, on the one hand, it is a way in which women contribute to the economic development of their countries and, on the other, because work is a major source of income which gives access to education, culture, power and other factors influencing social status. Many modernisation theorists argued that industrialisation would open up opportunities for women, providing them with a greater variety of possible roles and increasing chances of social mobility and a less subservient existence. There is a good deal of evidence to show, however, that industrial capitalism places women on the periphery of the economy, and for women in a Third World country which is itself on the periphery of a world economy the situation is even more difficult. In a pre-industrial society women may have important occupations in handicrafts or in garden agriculture – jobs which can be done in the home. But industrial development takes remunerative work out of the home, making it more difficult for wives and mothers to participate. Women participate in poorer jobs and in the tertiary sectors, areas which have suffered the most from peripheral capitalist development.

The high rates of unemployment so characteristic of underdeveloped countries make women especially vulnerable. For example in Brazil, peripheral capitalist development is responsible not only for a lower level of participation of women in agriculture but also a lower level of integration of women in urban development. In 1872, 45.5 per cent of the workforce of the nation were women, but by 1920 this proportion had dropped to 15.3 per cent and by 1950 it had fallen even further to 11.3 per cent (Saffioti 1978). 'It is not industrialisation *per se* that creates underemployment and marginalisation, but an industrial finance capitalism and advanced technology (intensive in capital) which does not permit fundamental changes in the economic structure' (Vasquez de Miranda 1977: 273).

In the stage of early capitalist development the demand for labour in a previously agricultural economy is high. But in a situation where industrial growth is conditioned by the needs of monopoly capitalism, as happened in many underdeveloped countries, this stage may be virtually omitted. Because of multinationals, industrial production means little improvement in levels of employment. Therefore, opportunities for men and women, though especially the latter, are limited to traditional and backward sectors such as subsistence agriculture and domestic service. The latter has become the major occupation for women, for about one-quarter of Latin America's female workforce are maids – 25.5 per cent of the female labour force in Bogotá in 1985 were domestic servants (López and Pollack 1989) and 32 per cent of the total Brazilian female workforce are live-in maids (Filet-Abreu de Souza 1980). This is particularly the case for women of colour. In Brazil domestic service accounts for 56 per cent of jobs Black women fill compared to 24 per cent of the jobs occupied by white women (Caipora Women's Group 1993: 12).

Nash argues that the position of women has deteriorated despite development activities because of the concentration on production for profit rather than for the welfare of the population in many Third World countries. This leads to a so-called rationalisation of economic activity which implies the increasing use of technology at the expense of human labour (Nash 1977).

One of the reasons why women are more vulnerable than men with the introduction of highly technologised industry is that female workers are considered inappropriate for mechanical industry. In Guatemala, for instance, changes in the character of the manufacturing sector influenced the participation rates of women in the labour market. The industrial censuses of 1946 and 1965 showed a decline in the female workforce from 22 per cent to 18 per cent. The largest declines were in textiles, tobacco, chemicals, rubber, paper and food. For male workers, however, employment was rapidly increasing in the area of electrical appliances, transportation and furniture which first appear in the 1965 census, as well as metal products and some of the areas traditionally open to women – chemicals, paper and rubber (Chinchilla 1977).

Similarly, Bolivian women lost out to men in the mechanisation of the mines. From the early period of tin mining to the 1940s women were often concentrators of minerals. When new methods of concentrating minerals were introduced, large numbers of female workers were replaced by male workers in the sink and float plans. Since 1960, with the complete mechanisation of the mines, no women have worked as metal concentrators (Nash 1977).

As a result, up until the late 1970s the rate of labour-force participation by women remained fairly stable, hovering at around 20 per cent. However, by the 1980s the picture was rather different. The period from 1960 to 1985 witnessed a drop in male participation rates in the labour force and an increase in female participation (see Table 5.4).

The two major causes of the drop in male rates are an increase in education delaying the age of entry of young men, and the extension of Social Security coverage, which has allowed more men to retire at a younger age on pensions (Arriagada 1990). It must be remembered that in the 1980s this takes place within the context of a declining economy.

In the case of female participation, there are both positive and negative factors at work. The logic of choice makes it possible for women to benefit from improved levels of education and work for personal fulfilment. In Brazil the rates of participation in the workforce of women with university education are 66 per cent for married and 77 per cent for single women (Filet-Arbreu de Souza 1980). At the same time, the logic of determination, which affects a greater number of women, forces low-income women out to waged work in the face of poverty. National indebtedness and structural adjustment policies carried out in Latin America during the 1980s led to declining incomes and greater impoverishment within the popular classes. During the crisis years in Mexico (1980–86), twice as many women entered or re-entered the labour force in Leon as those who had left and in Puerto Vallarta the net rise was about 20 per cent (Chant 1991).

Table 5.4 Participation in work by sex

Country	Men		Women	
	1960	1985	1960	1985
Argentina	78.3	67.1	21.4	24.7
Bolivia	80.4	70.9	33.2	21.5
Brazil	77.9	71.8	16.8	26.6
Colombia	75.5	67.3	17.6	19.2
Costa Rica	79.3	73.5	15.0	20.6
Cuba	72.7	64.0	13.9	29.6
Chile	72.5	65.2	19.7	24.4
Ecuador	82.1	69.2	17.3	16.6
El Salvador	80.7	72.9	16.5	24.3
Guatemala	82.0	71.7	12.0	12.9
Haiti	84.0	72.9	72.1	52.2
Honduras	82.7	74.5	13.7	15.6
Mexico	72.5	68.1	14.3	25.0
Nicaragua	80.5	70.8	17.3	21.3
Panama	75.8	67.1	20.2	25.4
Paraguay	78.5	75.5	21.3	19.5
Peru	73.1	66.5	20.4	21.4
Dominican Republic	75.9	70.7	9.3	11.3
Uruguay	74.3	67.6	24.1	28.2
Venezuela	77.1	68.4	17.2	25.3

Source: Arriagada, *CEPAL Review* **40**, April 1990.

These developments have meant a change in the structure of female occupations. As with male employment, there has been a distinct shift away from the agrarian sector and a process of tertiarisation, leading to the majority of women working in the service sector.

Some sectors of the economy have been characterised by a feminisation of work, i.e. greater percentages of women are found in each occupational group. The most outstanding is that of domestic service. An ECLAC study of six countries (Argentina, Brazil, Chile, Ecuador, Panama and Uruguay) shows that the proportion of maids and laundry people who are women ranges from 89.9 per cent in Ecuador to 98.9 per cent in Uruguay (Arriagada 1990). This varies considerably from other LDCs where domestic servants are both male and female. Most maids in Latin America live in, with the result that they can be compelled to work very long hours for very little pay. This is the major form of work for female rural–urban migrants, and although it does offer some source of income, it also places the maid in an exploitative position.

Feminisation has also occurred in the categories of nursing and teaching. In the ECLAC study, the majority of workers in these areas were women, with rates ranging from 54.9 per cent (professors and teachers in Ecuador) to 84.7 per cent (professors and teachers in Argentina). These occupations require an extension of the caring wife–mother role, making them apparently appropriate for feminisation (Arriagada 1990).

Some large firms have demonstrated a preference for female workers, often on the grounds of their supposed nimble fingers, though in reality usually because they can be paid less and are less likely to unionise and form any opposition to management. Firms that have employed women in this way are, for example, the American concerns, the *maquiladoras* that have been set up in Northern Mexico near the US border. They employ women for long hours and often in poor conditions in the production of textiles, pharmaceutical and a variety of other items. Positions such as directors, managers, administrators and owners remain firmly in male hands.

In general women have not benefited much in the job market from capitalist industrialisation, nor have they gained much influence in society outside the family through political channels. The political activity of women is limited and, in a few cases where they do achieve positions of some power, these positions are often linked to the traditional female role. Elsa Chaney, in her study of women active in government in Peru and Chile in the late 1960s, found that they were overwhelmingly engaged in feminine stereotyped tasks related to education, health, social welfare and cultural fields. Out of the 167 women interviewed, nearly 70 per cent did stereotyped female work and/or held appointments in agencies carrying out tasks considered of particular concern to women. As a result, she epitomised the successful political woman's role as that of 'supermother' (Chaney 1979).

If capitalist industrialisation has done little to improve the position of women, is there any evidence to suggest that they have fared better with alternative paths of development? Both Nicaragua and Cuba have seen significant changes for women, but the Cuban example is particularly instructive, since there has been time for new gender roles to develop.

As the Cuban revolution was dedicated to reducing inequalities of all kinds, it was only a year after the revolution that the Federation of Cuban Women (FMC) was set up to tackle the problems facing women. Since prior to the revolution the majority of women had very low levels of education and therefore few opportunities open to them, the FMC embarked on an educational programme and a rehabilitation scheme for the many women who found themselves in degrading and unrewarding situations. To this end, schools were set up for peasant women and prostitutes (who were numerous since Havana had previously been a notorious entertainment centre for North Americans), to give them an education and to provide them with some skills and training ready for the job market. It must be remembered, too, that the early years of the revolution saw a great demand for labour, in order to rebuild the economy and, also, to fill the immense gap left by the departing middle class. As industrialisation takes work out of the home, women, if they are to be encouraged to join the labour force, must be assisted in child and home care. The state, prompted by the FMC, has taken numerous measures to encourage women in this way and has set up nursery schools, crèches, workers' canteens, automatic laundries, the provision of medical facilities and a system of grants. Laws were passed to make it easier for women to participate fully in the development of their country as well as maintaining their family life. Perhaps the most radical of these is the Family Code, adopted in February 1975. This is one of the most progressive pieces of legislation in the world in attempting to establish equality between men and women as the rule in marriage. According to this law, both partners in a marriage have the duty to help each other, must share the housework and have the same obligations and duties concerning the raising and education of their children. The code particularly guarantees the participation of both men and women in the economic, cultural and political order of the country (Latin America and Caribbean Women's Collective 1980). Nor is the law an arid statement of ideals, for defaulters may be taken to court by their partners.

There is no doubt that the FMC which, by 1980, represented over 80 per cent of Cuban women, has done much to improve opportunities for women and, above all, to give them dignity and respect in society. They provide a very active participation in the many voluntary groups supporting the revolution and contribute much more than previously to development through participation in the labour force.

The main area of weakness lies in the political arena. When it comes to the obvious source of power, gender differences seem to be similar in socialist Cuba to capitalist Latin America. Most full-time political representatives in Cuba are men, with very few women standing for election and even fewer succeeding. This problem became one of national concern when Fidel Castro commented on the elections of the Popular Power Assembly in Matanzas Province. In this case women represented 7.6 per cent of the candidates proposed and only 3 per cent of those elected. Investigations followed which revealed various reasons for this state of affairs. Lack of time was one factor, for many women now had to work outside the home as well as bearing family responsibilities. Some felt that women lacked the cultural and political training necessary. *Machismo* was still very evident, for many people felt that men were better suited than women to lead and, in some cases, women had been forbidden by male relatives to enter politics.

Cuba has gone a long way to reducing gender inequalities, though power relations are clearly favour men, a fact of which all Cubans, including their leaders, are very aware. In 1974 Fidel Castro said, in his closing speech to the Second Congress of the Federation of Cuban Women, 'We live in a socialist country, we made our revolution sixteen years ago, but can we really say that Cuban women have in practice gained equal rights with men and that they are fully integrated into Cuban society?' (Latin American and Caribbean Women's Collective 1980: 96)

RACE RELATIONS

Another area of Cuban society in which the issue of equality within the realm of social relations has had to be addressed is that of race relations. The revolution moved rapidly to counter institutionalised racism. On 22 March 1959 Fidel Castro made a well-known proclamation against discrimination, in which he denounced racial discrimination and prejudice as 'anti-nation'. In the 1960s, pronouncements on race took their cue from Martí: man should be judged according to merit and not race, and above all the enemy should not divide and thereby rule (Sarduy and Stubbs 1993). The revolution clearly ironed out many excesses of racial discrimination and eliminated its overt manifestations. The Cuban government has, however, been more concerned with the general redistribution of goods and services rather than the singling out of colour for particular attention. Many Black Cubans certainly did stand to benefit from programmes of the revolutionary government aimed at the poor.

Black people in Cuba continue, however, to dominate in the older, poor neighbourhoods. They continue to be marginalised in the sphere of personal, social and cultural relations, despite the evident racial mix within the country.

Thirty years is obviously a relatively short period of time for deep-seated attitudes and values to change. The intermingling of three different stocks (Indigenous American, European Caucasoid and Black African) throughout Latin America led to a complex racial situation, often referred to as the colour class system. The implication is that generations of miscegenation make it difficult to define racial boundaries; nevertheless, there is a clear correlation between skin colour and racial features on the one hand and class structure on the other. In Brazil this composition derives from the social organisation of the colonial system, and the social changes since independence and abolition of slavery. The status of African or Indian women impregnated by white masters varied: some became domestics, others concubines and a very small number lawful wives. As regards their children, at first the racial intermingling of people was clearly denoted, with special terms for those with one-quarter 'white blood' and those with one-eighth. Many illegitimate children inherited small portions of estates or were promoted from slavery to petty bureaucratic positions in areas where there was an absence of Spanish colonisers. These groups formed a kind of middle class with multiple layers; it also acted as a buffer between the ruling and the most exploited groups.

Gilberto Freyre (1946) is responsible for the generalised view that African slavery in Latin America was a gentler institution than in the Northern hemisphere. Freyre emphasised the intimate links between masters and slaves, basing his analysis on the Brazilian *fazenda*. Other historians developed and gave substance to this idea, and from here the myth of 'racial democracy' was born. 'Racial democracy', which applies to Latin America, is in effect the process of 'whitening' through miscegenation (Radcliffe and Westwood 1993: 6).

> The emergence of a large mixed intermediate group . . . has established the myth of a Latin American 'racial democracy' based on the predominance of the *mestizo* and the mulatto and in which racial marks are no barrier to marriage and social mobility. It is important to recognise, however, that the mechanisms of racial and social vertical mobility that exist in Latin American societies draw their dynamic from an attempt to escape a blackness that has been and continues to be negatively evaluated, and thus to whiten oneself and eventually the population as a whole.
>
> (Wade 1986: 16)

Racism in Andean and Middle American countries follows a similar trajectory, whereby those groups with perceived Indigenous heritage have been viewed negatively throughout post-contact history (Radcliffe and Westwood 1993; Bourricaud 1975; van den Berghe and Primov 1977). In Mexico, with its rich anthropological tradition, the *indigenistas*, stressing the cultural distinctiveness of the Indian communities, felt this to be an impediment to modernisation and incorporation in the wider society. Since they believed in a benevolent national society, they considered integration in the form of a move from Indian to *mestizo* to be upward social mobility.

All three racial stocks, the Indigenous people of the continent, the European Caucasoid, who came to dominate in the New World and the Black African, have contributed to the cultural heritage of the continent. The contribution of the non-European groups varies in strength and kind across South and Middle America, according to the historical incidence and development of the different ethnic groups (see Fig. 5.1).

The acceptance of doctrines such as racial democracy have led to a denial of a racism which very clearly exists in Cuba and which is worse in other Latin American countries. Racism in Brazil, for example, while having long been dismissed as a preoccupation of foreigners means that around 90 per cent of inhabitants of the *favelas*, the majority of psychiatric patients, of prison inmates, of the people who lack food, clothing and education are black (CWG 1993: 12).

In Cuba the trend of dismissing racism as a factor of oppression due to doctrines of racial democracy was compounded when political–economic class identity became the predominate mode of analysis in post-revolutionary society. Any ethnic referents were therefore made invisible. Invisibility is a wider concept which concerns a stage of racism which Michelle Wallace sees as the final and most difficult to combat (Wallace 1990). It concerns an inability to recognise cultural difference and to understand that people of different ethnic backgrounds can all be equal in their differences. With the fall of the Berlin Wall there has been an inevitable and wider recognition of the weaknesses of Marxist theory, such as the serious obstacles which it proposes to the analysis of theoretical categories beyond nation and class.

This has gained some practical recognition in Cuba. In April 1992, the 6th Congress of the Communist youth debated openly many problems, including prejudice against young people and their life-styles. Demographically and socioculturally, the country is far less white after three decades of revolution than it was in 1959. The initial exodus in the 1960s was predominantly Hispanic and monied (Sarduy and Stubbs 1993: 10). One striking aspect of the congress was the composition of young people taking part – many were black and many were women. The background to such advances, others include the heavy racial symbolism woven into the 4th Congress of the Communist Party, held in October 1991 (see Sarduy and Stubbs 1993), has been a growing respectability of African culture and religion due, in part, to the visits of several dignitaries from various African countries. In 1987 two major traditional African leaders visited Cuba, the first was Asantehene of Ghana, King of the Ashanti, and the second, the Ooni of Ife, sacred capital of the Yoruba. Both, but especially the Yoruba, are peoples from whom many African Cubans are descended. In July 1991 Nelson Mandela also visited Cuba.

Such revelations in Cuban society led to a theoretical body which explores mechanisms of oppression and argues that it is of utmost importance to include theoretical consideration of, for example, both 'race' and gender. Postmodern research has allowed for the study of gender and race, not necessarily in isolation, but also in complex combinations.

A group affected by a combination of repressive social relations within the realms of gender, race and often class are women of colour. Study of their particular situation is recommended by Bell Hooks (1982) and Hull, Scott and Smith (1982), and also by Michelle Wallace (1990), who points out that women of colour, who are so often thought of as too small a minority to warrant conceptualisation, in fact represent the poor majority in a capitalised world system.

'The condition of women of colour cannot be separated from the colonial experience since the basic paradigm of power relationships (in Latin America) was established during the era of imperialist expansion by Europe into the New World' (Belén and Bose 1993: 56). This view is also taken up by Black women North American theorists, who have also had a dominant culture imposed upon them which prescribes the current patriarchal form of gender relations.

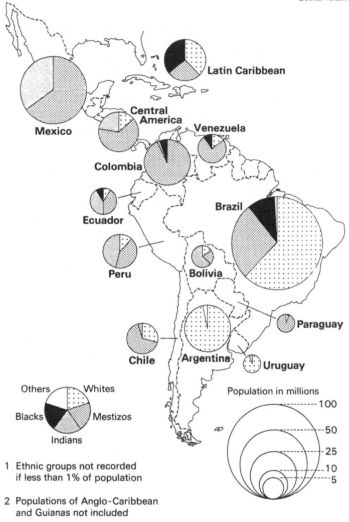

Fig. 5.1   Latin America: ethnic groups. (Cubitt D. and T., 1980)

CONCLUSION

The manipulation of the web of personal relations has an ambiguous role for, on the one hand, it serves to ease difficulties and build bridges across social divides but, on the other, it helps to maintain the status quo and thus the inequalities in the social structure. Resorting to personal ties to survive or advance is a mechanism which has been used to some effect over the generations, but which prevents people turning to organisations or political activity which might in the long run prove more effective in solving their problems. Inequalities between patron and client, peasant families and landowning families, men and women and Blacks, Indians and whites continue.

Despite rapid changes in a number of areas, the family remains an important and key institution. In times of change it is a source of support and strength, but also adapts to the new economic environment. With household survival strategies, the family calls on its resources in much the same way as the rural extended family.

Although women are participating more in development, through improved educational opportunities and greater involvement in paid labour, the benefits are still not commensurate with their role. The growing participation of women in social movements may lead to increased empowerment, which could provide the opportunity to address this imbalance of activity and rewards.

The conceptualisation of women as a last colony, (Mies, Bernholdt-Thompson and Werloff 1988) has provided a valuable interpretive model for feminist research on Third World populations. This framework underscores the convergence of race, class and gender without dividing these issues, and recognises one complex but coherent system of oppression. It allows us to see that patterns of sexism are integral to the simultaneous layer of oppression that Third World men and women share as a result of the colonising experience.

FURTHER READING

Chant, S. (1991) *Women and survival in Mexican cities.* Manchester University Press, Manchester and New York.

Jelin, E. (1991) *Family, household and gender relations in Latin America.* Kegan Paul/UNESCO, London.

Jaquette, J. (1991) *The woman's movement in Latin America, feminism and the transition to democracy.* Unwin Hyman, London, Boston.

Molyneux, M. (1985) Mobilization without emancipation? Women's interests, state and revolution in Nicaragua, in **Slater, D.** (ed.) *New social movements and the state in Latin America.* CEDLA, Amsterdam.

Radcliffe, S. and Westwood, S. (1993) *VIVA women and popular protest in Latin America.* Routledge, London.

Rothstein, F. (1979) The class basis of patron–client relations. *Latin American perspectives,* **21**, Vol. 6 (2).

6

Changing rural society

Although Latin America is becoming increasingly urban, rural society still represents a very sizeable sector. At the beginning of the 1990s about a quarter of the total labour-force was employed in agriculture. The farmers of the region are predominantly peasants who depend mainly on family labour, as nearly four-fifths of the agricultural sector are small-holders, cultivating one-fifth of all farmland.

The characteristic feature of this sector is its heterogeneity, brought about by the changes of the last few decades. Some countries such as Mexico, Brazil, Argentina and Colombia, are at the forefront of agricultural development. The vast irrigated areas of Mexico's north-west have produced a highly modernised fruit and vegetable industry. In the south of Brazil there are multi-million-dollar soybean-processing plants owned by multinationals and surrounded by large-scale mechanised farms. In other areas, like the Peruvian and Bolivian Andes, traditional farming techniques such as the digging stick, machete and hoe are used. Countries vary as to the proportion of the farm labour-force taken up by the peasantry. In Bolivia, the proportion is 90 per cent; in Brazil, Ecuador, Panama, Peru and Venezuela it fluctuates between 70 per cent and 80 per cent; in Guatemala it is 60 per cent; in Colombia, El Salvador and Mexico 50 per cent; and in Argentina, Costa Rica, Chile and Uruguay less than 50 per cent (López Cordovez 1982).

This chapter examines the way that change has taken place, in particular in relation to the social structure in terms of changing social relations of production, and the contribution to this process of change made by different social groups, such as political élites and the peasantry.

The changes that take place vary and result from three different kinds of processes.

The first of these is the penetration of capitalism into rural areas, which brings about the spread of the commercial nexus, usually through the market system. This generates a number of economic changes which, because of the overlap of social, economic, political and religious systems in rural communities, have extensive social repercussions. Secondly, some change may be organised from above in order to encourage agricultural productivity and curb social injustice. These are agrarian reform programmes, planned at government level to alter land distribution and modernise traditional agricultural methods, and also policies concerning food production and distribution. Thirdly, peasant mobilisation may be a source of change. Although the peasantry have not often been prompted to rebellion, they have played a major role in a national revolution in two Latin American countries: Mexico and Cuba.

The peasant economy, which is affected by these processes, can best be defined in Stavenhagen's words:

> The peasant economy can be defined quite simply as that form of farm production (and associated activities) in which the producer and his family till the land themselves, generally utilising their own means of production (tools and instruments), with the object of directly satisfying their basic needs, although for a number of reasons they may find themselves required to sell a part of their produce on the market in order to obtain goods which they do not produce.
>
> (Stavenhagen 1978: 31)

PENETRATION OF CAPITALISM INTO RURAL AREAS

Andrew Pearse describes this process as the incorporative drive of national market systems, and he argues that it is the major cause of rural change (Pearse 1975). He does not deny internally generated change, which might occur from natural factors such as drought or population growth, but suggests that the latter forces are more important in pre-industrial conditions. The incorporative drive refers to national level forces which emanate from an urban centre and gradually involve more peripheral areas in a national way of life. It includes three main elements: (1) the spread of the market system; (2) the spread of a communications system; and (3) the penetration of national institutions.

The spread of the market system brings the peasant into contact with a nation-wide network of distribution. Business is now done through a middleman, an outsider with whom he, initially at any rate, does not have a personal longstanding relationship. The market makes it possible for the peasant to grow and sell cash crops, rather than just maintaining a subsistence economy, and therefore to produce a surplus. The penetration of the national capitalist system also makes available the consumer goods which can be purchased with the income from the surplus. The market system holds out opportunities for the peasant, but it also makes his situation more precarious. As the peasant moves from subsistence agriculture to the production of cash crops, he loses the security that a subsistence plot provides, i.e. a basic food supply. If the cash crops fail for any reason, he has nothing to fall back on. The peasant is also vulnerable in the market, for he has to operate within a new system of values: one that emphasises competition, astuteness and individuality. He is unfamiliar with national levels of price-fixing and laws of supply and demand.

At the same time, improvements in communications raise aspirations and hopes. Radio, television and films project images of sophisticated living that are often presented as being available to all who work hard enough. Consumer goals which were never contemplated by the peasantry in the past, now appear on the screen in their own homes and villages. Better roads and more efficient and cheaper forms of transport have facilitated contact between town and country.

The penetration of national institutions seeks to incorporate the peasantry into a variety of local segments of national organisations, such as schools, church groups, political parties, peasant unions and cooperatives. These are often sponsored by the urban middle class, with the result that there may be some incongruence between these programmes and the needs of country people.

Lenin contrasted two types of development that can result from the penetration of capitalism in agriculture: the Kulak (or American) path and the Junker path. With the Kulak path, some members of the peasantry adapt successfully to the market system, thus becoming relatively wealthy, while other peasants only become impoverished, so creating internal social differentiation within the peasantry. The wealthy become capitalist farmers, often employing the poor whose impoverishment leads to the loss of their land. The Junker path involves the increasing proletarianisation of the peasantry, who lose their plots to encroaching landlords (for a more detailed discussion of Kulak and Junker paths, see de Janvry 1981).

Pearse identifies the same process as the Kulak path when he shows how the incorporative drive draws out what he calls the progressive element among the peasantry (Pearse 1975). A small number are able to manipulate the new situation because they are able to get credit, to establish good terms of trade with middlemen and generally to come to terms with the different conditions. The majority of the peasantry, however, have neither the resources nor the social advantages necessary to become entrepreneurs. Because of debts they are often forced to sell their land and therefore lose the means of livelihood that their family has known and understood over the generations. Security is lost and the traditional social networks have broken up in favour of the commercial nexus.

The Kulak path has occurred in many different parts of Latin America. Changes that have taken place in the Highland Zapotec region of Oaxaca, Mexico, are a good example. Kate Young's data from the two villages of Copa Bitoo and Telana reveal how social differentiation can take place (Young 1978). In the nineteenth century, both villages were self-sufficient, mostly in maize, the land was owned communally and there was no internal stratification. Prestige was allotted according to the principles of the ritual cycle. The little money that was needed to pay state taxes and church tithes and to fund festivals was obtained from the sale of cotton lengths, which were made by the women. In the 1860s the supply of cotton dried up because the Mexican cotton growers had started to replace the North Americans (who had become involved in their Civil War) as suppliers of cotton for England. As Oaxacan peasants needed some source of cash, the men started seasonal migration, which was disliked because the plantations had a reputation for malaria and poor living and working conditions.

At this point the two villages started on different paths. In Telana, coffee cultivation was introduced by the local *cacique*. Coffee could be grown quickly, cheaply and easily and sold for cash. Gradually coffee came to replace maize as the main agricultural produce of the community and foodstuffs were bought with surplus cash. Specialisation was underway and since households were now in competition with each other, reciprocal labour agreements disappeared and each household became more dependent on its own resources. During the turbulent years of the revolution, coffee production declined, but in the 1930s new coffee wholesalers entered the area and began to buy for cash. Telana now became involved not only in the national market economy but also in the international economy, as these wholesalers were buying for export. During the period from 1938 to 1954 the world market price of coffee rose twenty-two-fold, so Telana prospered from these boom years. The community was now devoted to coffee growing, subsistence plots had disappeared and internal differentiation was taking place. Those who did well out of coffee growing started employing labourers and consolidated their control over the best land. The women, who had lost their craft specialisation, had few opportunities to develop any new specialisms and so lost the independence and prestige they had had from making cotton lengths.

After 1957 coffee prices fell drastically, with the result that all the villagers suffered. They had no subsistence plots to fall back on, some tried reviving traditional handicrafts but these could not compete with the manufactured goods that had been made available by the market system. The wealthier survived because they had control of tracts of good land, but the poorer villagers had to sell the little land they had to pay off debts, and were compelled to work as landless labourers for the rich. The 1970s saw considerable migration from Telana, as desperate peasants went in search of work.

The peasantry of Telana had become involved in international trade, with the result that they were affected by world booms and slumps. Some had fared considerably better than others, leading to a differentiated society. Kate Young found that three classes had developed on the basis of wealth, land ownership and relation to the means of production. Firstly, there were the rich, who owned farms and employed labour in both agriculture and the home and who made up 20 per cent of the population. The middle 30 per cent owned land but did not employ labour, as they relied on family members. The poor were the landless who provided labour for the rich.

Copa Bitoo, although internally differentiated, had not developed such a marked stratification system. Copa Bitoo had moved away from a sub-sistence economy, but was producing fruit and nuts for the local market, not for export. Here, similar classes were emerging but only about 10 per cent were able to employ wage labour regularly and only 20 per cent provided it regularly. The bulk of the villagers (the remaining 70 per cent) fell into the middle category, as they owned land but did not employ others to work on it.

Both communities demonstrate the way that involvement in the market system creates differentiation, though this is greater where bigger fluctuations take place. Participation in local markets and national capitalism generates differences, but insertion in the international system accentuates the peaks and troughs of economic cycles, which means a greater gap between rich and poor in peasant communities.

The categorisation of rural change into the two alternative paths of Kulak and Junker is schematically useful, but does not exhaust the possible forms of agricultural development. Roxborough demonstrates how both paths have been combined in the agricultural history of Chile (Roxborough 1979). The process of proletarianisation itself is a more complex issue than might at first be supposed. Proletarianisation implies an increase in the number of wage labourers, which can be demonstrated by statistics, but figures also show that in some parts of Latin America the numbers of the peasantry are increasing. Between 1970 and 1981 the agricultural population of Latin America increased by eight million persons, of whom five million were peasant farmers: four million are described as producers, and one million as landless workers, who must therefore seek work in wage-earning occupations (ECLAC/FAO 1985). At the same time, as the next chapter shows, many of the rural population have been migrating to urban areas in the hope of joining the labour force there. Is it the case that the rural sector is undergoing a process of proletarianisation or peasantisation? (See Goodman and Redclift 1981 for a good discussion of this debate.)

PROLETARIANISATION OR PEASANTISATION?

Marx originally argued that the growth of large-scale industry led to the demise of the peasantry and the development of the proletariat. His argument is that the penetration of capitalism into the countryside leads to the articulation of the capitalist mode with the subsistence economy, which in turn leads to the diversification of the village economy, with wage migration becoming an increasing part of rural life. Small peasant farmers can not compete with capitalist concerns, with the result that they often lose their land to them and end up working for them or migrating to the towns in search of wage labour. Modernisation theorists too thought that the peasantry, through the diffusion of modern ideas and consumer goods, would develop out of existence. Some Latin American social scientists working on the topic have suggested a two-stage process, in which the penetration of capitalism initially leads to an increase in wage-earning workers, but this is followed by a stage of the intensification of capitalism, characterised by deproletarianisation and semi-proletarianisation (Miró and Rodríguez 1982). It is the argument of this section of the chapter that the deepening of capitalism in the countryside has led to a heterogeneous pattern of labour relations, according to area and nature of commercial expansion.

Proletarianisation has occurred in Latin America because the peasantry have been dispossessed of their land or because their land has become insufficient to support a family. Dispossession has been encouraged by the transformation of the estate to a modern commercial farm. Traditionally the landowner did not need to cultivate the land intensively to provide sufficient to maintain prestige and a very good life-style. Today, with greater opportunities for profits from export and the advantages of modern transport and mechanisation, the landowner exploits his property to the full to maximise production and profits. In this situation it is often cheaper to hire labour on the open market when needed than to maintain workers and their families permanently on the estate. For this reason many landowners have followed the Junker path, taking over the land of sharecroppers and *colonos* and compelling them to sell their labour on the open market. Small peasant farmers who own their own plots have often been compelled to sell land through poverty and debts. As land is passed on over the generations, it gradually ceases to be a viable unit.

The process of proletarianisation has also received some impetus from the spread of agribusiness in the region. Agribusiness refers not just to large-scale rural enterprises, but to concerns which have an integrated system of production and distribution. Agricultural produce is grown, processed and distributed within the one company, although these activities may, and usually do, take place in different parts of the world. The influence of agribusiness is felt throughout the Third World, though

North American firms have found a major outlet in Latin America. The areas of agro-industry in which transnational corporations and other private foreign investors in Latin America are most active are branches of milling, oils and fats, chocolates and sweets, and meats and dairy products. This gives them a distinct influence on the development of agriculture and the food economy of the region.

Large-scale commercialised farming, that is agribusiness and domestic agricultural enterprises, have to compete with other big corporations for investment capital and markets, and must therefore make profits as efficiently as possible. To this end, commercialised farming emphasises different forms of labour utilisation according to the needs of the company in different areas. In some cases, firms (such as Nestlé in regions of Brazil), having purchased *haciendas*, are now running them with modern agricultural techniques. In other parts of Brazil giant corporations exist alongside small peasant producers with whom they have developed a complex interdependence, based on the supply of goods rather than labour. In several countries the second stage of the Miró and Rodríguez thesis is in operation. For example, the Banana Company of Costa Rica, a subsidiary of United Brands, has been converting plantations to African palm production, an enterprise which has a guaranteed internal market as Costa Rica imports nearly all its vegetable oil. This economic activity requires less than half the labour needed in banana cultivation and therefore must result in the loss of a number of jobs (LAWR 1983). In some areas mechanisation is replacing labour, although in the countryside this has not gone as far as it has on North American farms. Some harvesting machines are now being used in Brazilian sugar plantations and Mexican sugar and cotton fields. The shift from labour-intensive coffee growing to capital-intensive soybean production in Brazil created unemployment for tens of thousands of workers in the early 1960s (Burbach and Flynn 1980).

The form of social relations of production which is on the increase in the region is semi-proletarianisation. This is due to the seasonal nature of so much agricultural work, which makes it cheaper for the company to employ labourers during the planting or harvesting rather than maintain a labour force throughout the year. Seasonal workers do not have the security of full-time employment, since they must seek other means of support for part of the year. The pay is very low, as cheap labour is part of the incentive for entrepreneurial farmers, nor do they receive even the minimum security benefits of full-time work. Agribusiness that operates in sectors marked by seasonal fluctuations has been prone to this sort of employment. Del Monte employed in 1976 about 1750 workers in Irapuato, Mexico, but only 120 were permanently employed. Ninety per cent of the remaining seasonal workers worked for four to six months a year for a low wage (Burbach and Flynn 1980).

Semi-proletarianisation takes a variety of forms. A rather unusual

133

example are the *boias-frías* in Brazil. These labourers, who have no access to land ownership, live in urban areas but are unable to find work there. They therefore travel daily from town to country to do seasonal work as agricultural labourers. Their name is derived from their food, which they are obliged to take with them and eat cold. Semi-proletarianisation is closely linked to temporary intra-rural migration, as numbers of peasants move to a particular area for seasonal work and then move back to their small plots of land or on to another area, where a different crop has reached the planting or harvesting stage. This is particularly apparent in Central America, where it is estimated that seasonal migrants make up 70 per cent of the labour force employed in agriculture. In El Salvador, about half the active agricultural population are employed for less than six months of the year (Miró and Rodríguez 1982). Guatemala's large seasonal migrant population, who work on the coffee and cotton plantations for part of the year, sometimes migrate over the border to the plantations in southern Mexico. Intra-rural migration is also evident in South America. On the Bolivian *altiplano* 1.2 persons per peasant family migrate temporarily to seek work (Ortega 1982). In areas such as northern Argentina, central Chile, Peru, various regions of Brazil and Mexico this is becoming a basic element in the subsistence of the economy.

The intensification of capitalist farming has not meant the disintegration of the peasantry, for there is evidence to show increasing numbers of peasant farmers in many regions. Census data for a group of seven countries, Brazil, Chile, Colombia, El Salvador, Honduras, Peru and Venezuela, for the 1960s and 1970s show that the number of agricultural units of less than 20 hectares in these countries rose from 4.7 million to 6.5 million, that is by 38.5 per cent (Ortega 1982). This expansion is the result of several factors. Firstly, subdivision, because inheritance is shared between siblings, takes place throughout the region. Secondly, agrarian reform in some areas such as Peru and Chile (in the early 1970s) has had the effect of dividing large expropriated landholdings among the peasantry. Thirdly, new territories have been cultivated, thus extending the agricultural land available, as for example the colonisation of new lands in Amazonia. Between the 1950s and early 1970s, 140 million hectares in the region were brought under cultivation for the first time (Ortega 1982).

How can the existence of these two apparently contradictory trends, proletarianisation and peasantisation, be explained? First there is an obvious demographic point that has to be made: the population has been increasing so fast in most Latin American countries that despite outmigration the numbers of the peasantry are still on the increase. But there are more fundamental factors to which attention must be drawn. Industries in the LDCs have demonstrated a structural incapacity to absorb all the migrants and all those in search of work. Neither industry nor farming can absorb the population of working age. As a result, many of the peasantry maintain a subsistence plot but, since this is usually insufficient to support a family, the male members of the family have to seek seasonal labouring work, leaving the women to maintain the family farm. In Mexico, where half the number of farm units are infra-subsistence, i.e. not big enough to support a peasant family, the peasant must find other forms of income (Stavenhagen 1974). The small farmer can combine the seasonal work offered by agribusinesses and urban enterprises with his family farm, and indeed must do this to survive. For the employer, a part-time workforce means a cheap labour supply which contributes to capital accumulation.

In some cases the link between the peasant sector and commercial farming may rest on the sale of goods rather than labour, as in the following example taken from the State of São Paulo.

> The almost absolute dependence of the small and medium-sized landholders on large-scale capitalist enterprise is illustrated by the case of tea, where the agro-industries possess their own *haciendas* and the organisation of labour is completely of the wage-earner type. However, these agro-industries also deal with family or independent units of production, to which they supply fertilisers and other production inputs. These same enterprises send their trucks for transporting merchandise during the harvest periods, and the classification of tea leaves for quality is also done by the enterprise, without the participation of the small producers, who are paid in accordance with this classification. These small farmers may be tenant farmers, sharecroppers or owners of small farms.
>
> (Miró and Rodríguez 1982: 57)

This continuation of the peasant sector alongside the development of industrialisation is a distinct aspect of capitalist development in the periphery and characterises much of rural development in the LDCs. Capitalist industrialisation in the periphery, often because of transnationals which can import sophisticated technology, has not been able fully to absorb all the labour available. Commercialised farms that do rely on labour frequently use seasonal labour. The subsistence economy, however, provides security for irregular workers. The peasant economy, therefore, reproduces the labour force for capitalism at a low cost because, due to the subsistence sector, men can be employed on an irregular basis. The peasant economy provides a reserve of cheap labour for capitalism and so contributes to capital accumulation. Moreover, subsistence plots produce goods at competitive prices by the working of long hours, using unpaid household labour and accepting low standards of subsistence. The conclusion of this is that 'the constant recreation of the peasant economy is functional for the capitalist system' (Stavenhagen 1978: 35). Thus the capitalist mode articulates with the peasant mode, with the latter playing a subordinate role and the former benefiting. The capitalist economy, while ensuring the continuance of the peasant sector, does not allow it to develop sufficiently to extricate itself from its subordinate role, for this would destroy its value for capitalism. This articulation, however, does generate its own contradictions. The resulting poverty of the peasantry, who must find work in two different sectors to survive, limits their spending power and therefore discourages the purchase of consumer goods. This inhibits the growth of a market which would stimulate local economic development and so promote the expansion of manufacturing firms.

AGRARIAN REFORM

Both international politics and the philosophy of the day gave rise to a number of agrarian reform programmes in the 1960s and 1970s. Prompted by the fear of another Cuba in the region and desires for economic growth, Latin American political élites came to an agreement with the USA over aid for reforms. Thus, the Alliance for Progress, which gave rise to many agrarian reform programmes, was set up in the early 1960s. At the same time, modernisation theory, still influential in the 1960s, suggested that the removal of obstacles to change was necessary for development and a major obstacle in Latin America was the *hacienda*. Modernisation implied the elimination of these feudal-like institutions and the modernising of agriculture, both socially and economically.

In social terms, the *hacienda* system represents a very unequal distribution of land, with a small number of people monopolising the bulk of cultivable land. In 1960, before agrarian reform got underway, there were 111 million rural inhabitants, of whom some 100 000 owned 65 per cent of total agricultural land (*International Labour Review*, July–August 1961). This pattern of land distribution reflects such great disparities and brings about so much poverty that it represents one of the major problems that agrarian reform has addressed.

The other problem is economic. The large estate system leads to an inefficient use of the two most abundant factors of production in Latin America: the land and the labour force. Traditionally, the large estates have concentrated on one crop, often for export. This leads to an underemployment of land and labour, because neither is being fully used throughout the year. Not only does this mean that the rural population is not being fully utilised in terms of labour power, it also means it is too poor to provide a substantial market for industry. Agrarian reform was needed to release the peasantry from traditional forms and low levels of remuneration in order to increase its purchasing power to provide a stimulus for the growth and expansion of local manufacturing firms.

Agrarian reform programmes have two basic aims, one of which is redistributive, to provide a more just pattern of landholding and alleviate poverty by giving the peasants more substantial plots of land. The other is productive, to provide more agricultural produce by more economic and efficient means of farming. In order to achieve these goals an agrarian reform programme should have three main elements. Firstly, a broader distribution of the ownership and control of land is necessary, but this on its own is not sufficient. Secondly, in order to make redistribution successful, both in economic terms and for the beneficiary, the implementation of modern, efficient and productive techniques is needed. It is not enough to enlarge a peasant's piece of land; he must have the wherewithal to cultivate it. Finally, to be successful, an agrarian reform programme should include comprehensive and realistic community development, including schools, health care, services for the protection of life and property, the construction of local roads and bridges, agricultural extension activities and farm credit facilities. A supportive infrastructure is needed to make the best use of the land.

Most Latin American countries have embarked on some sort of land reform programme, though these have varied considerably across the region in scope and depth. In countries like Colombia and Venezuela there has been some mild tinkering with the traditional agrarian system; in contrast Cuba has experienced a complete restructuring of its agrarian sector. Some countries, such as Mexico and Chile, have gone through periods of quite radical change, only to be followed by new political regimes, which have led to a reversal of policy.

In order to see what can be achieved by a successful agrarian reform policy, and indeed how these achievements can subsequently be reversed, it is worth examining the case of Chile. Agrarian reform was put into practice in Chile during the second half of the 1960s, with the passing of a law which made it possible to expropriate estates of over 80 hectares. The government began by taking over the most economically inefficient. The basic criteria of the reform programme were (1) that the peasants should gain ownership of the land they worked; (2) they should be encouraged to become full and active citizens of the nation; and (3) productivity should be improved at all levels of agriculture. Embodied in the reforms were the twin aims of social justice and increased productivity.

Once the land had been expropriated it was to be turned into cooperatives, but because the government felt it would be difficult for peasants who had had no previous experience of administering a large farm to move straight to cooperative farming, there was an intermediary stage. This, referred to as the land settlement, was run by the peasants but with the supervision of officials from the agrarian reform agency. The peasants, therefore, had the chance to get experience of administration and problem-solving, but at the same time had experts to assist and prevent mistakes of any magnitude. The intermediary stage was intended to last for three to five years. From 1964 to July 1970, agrarian reform involved the expropriation of 1408 properties, with about 30 000 beneficiaries (Kay 1978). Towards the end of the decade, with elections looming, land distribution decreased in favour of greater emphasis on raising agricultural yields.

The 1970 elections were won by a coalition of left-wing parties, led by a Marxist, Salvador Allende, with a radical programme which included a pledge for a total and speedy programme of agrarian reform. The new government decided to use the existing legal framework to save the time and possible difficulties that new laws might bring about, but with certain modifications. Expropriations were to be speeded up and the land settlements were to become new reform units, with wider criteria for membership, thus increasing the number of beneficiaries. Different types of reform unit emerged, but the most common, known as Peasant Communities, were not very different in their organisation from the previous land settlements. All the reform units offered their members security in the form of assured employment and access to a plot of land and full voting rights in the unit. The wider implications of agrarian reform were tackled in this period, for there was a redistribution of credit in favour of smallholders and the state took over some marketing networks, though this was primarily to get supplies to the urban consumer at low prices, rather than to assist agriculture (Castillo and Lehmann 1982).

Within two years the Popular Unity government had expropriated 3278 farms, that is, over twice as many estates as the previous government had done in six years (Castillo and Lehmann 1982). The main losers were the *hacendados*; those who gained were the permanent workers, who benefited from increases in the minimum rural wage and from secure employment. This meant a redistribution of income in the rural sector from the wealthy to the permanent workers, though it may be the case that there was a widening differential in incomes between permanent and temporary workers. The great achievement of this period was the eradication of the very large *hacienda* from the countryside. In 1973, according to official statistics, there were no privately owned properties of over 80 hectares in Chile. The economic results of the reforms are very difficult to judge because they were too short-lived for any economic stability to be achieved.

In 1973 Chile experienced a military coup in which Allende and thousands of his followers were killed, others fleeing into exile. The new government set about reversing previous policies, not least of which was agrarian reform. The agrarian reform units were dismantled, either by returning land to previous owners or distributing it to others. By 1979 just under 30 per cent of the land in the reformed sector had been restored to its former owners (Castillo and Lehmann 1982). The remaining land was to be given out in family units, in line with the government's ideology of private ownership. Since the remaining land was not sufficient to accommodate all the 75 000 beneficiaries of Chilean agrarian reform, the military regime started a selection process. A points system was introduced in which participation in the management of the agrarian reform units and in unions tended to count against people, whereas qualifications such as holding a degree in agronomy benefited them. The average size of each plot of land available rose, with the result that fewer people could benefit. The effect of this was that of the 75 000 original beneficiaries, 51 per cent did not receive any land. The military's policy brought about a greater disparity in landholdings, and by 1977 the large estate was back, since there were 1512 estates of over 80 hectares that year (Castillo and Lehmann 1982).

In economic terms, many of those who still had land in the years following 1973 were unable to farm it successfully. The institutions which had previously been set up to aid small farmers had their lending capacity cut considerably, making it difficult for many of the beneficiaries to find sufficient capital. Although until April 1980 it was illegal for beneficiaries to sell land, researchers found evidence of land being sold and also of farms that had been turned over to sharecropping. This suggests that many small farmers, unable to cultivate successfully, turned to the sale or renting of land. The farmers who are successful are those with medium-sized farms, who are growing for export. Agricultural and fishing exports increased fourfold in value from 1974–77. The produce is mainly fruit and luxury vegetables for upper-income groups. Export requires resources for capital investment, which excludes the small farmer, but the growing of fruit and vegetables does not necessitate very large tracts of land and, therefore, the medium-sized farmer can accommodate this. Chilean apples were on sale in British supermarkets in the early 1980s, despite the vast quantities of apples grown in Britain. The result in Chile is that poor, small farmers exist alongside medium-sized farms which are capital-intensive and produce for export, and a large number of the rural population have become landless and poverty-stricken.

The difficulty with many agrarian reform programmes has been achieving both social justice and economic growth. In Cuba, the former has been emphasised with marked success, while economic growth has been subject to fluctuations. Development there has been described by Lehmann as 'modernisation with fitful growth', because the expansion of the state apparatus has meant modernisation in terms of growing bureaucracy and an increase in formal, at the expense of clientelist and personal, relationships at work but at the same time agricultural productivity has been erratic (Lehmann 1982).

The predominant unit of production in rural Cuba is the state farm, although there exists, despite the nationalisation of privately owned businesses in the 1960s, a distinct private sector in the countryside, half of which lies in Oriente province, the home of the peasants who supported Castro in his bid for power. What the development of state farms has meant is the demise of the peasant plot based on unpaid family labour, the reduction of seasonal employment and a growth in proletarianisation. It is worth pointing out that before the revolution, Cuba, due to its sugar-based economy, had relatively fewer smallholders and more agricultural labourers than most Latin American countries. The revolution, through state control, has given agricultural labourers greater security and better wages by turning them into state employees. The success of collectivisation is revealed by the increasing number of cooperatives in the private sector. This mushrooming of the producers' cooperatives, often through the amalgamation of small farms has been welcomed by Cuban leaders because the cooperatives are generally well integrated into the revolutionary process.

Nicaragua, which has also had a socialist revolution, has a greater emphasis on the private sector in its agrarian policy. This sector plays a leading role, not only in controlling the greatest share of land, but also in producing three-fifths of the volume of agricultural production (Curtis 1983). Because of Nicaragua's dependence on agricultural produce and agro-exports, it has been important not to let production levels fall. As a result, because the properties that were expropriated were modern commercial farms, the government has maintained the structure of production, but, unlike the organisation prior to the revolution, the workforce now has a say in administration. Small farms in the public sector have been grouped together in cooperatives to gain all the benefits possible from size and concentration. At the same time, the social benefits in the few years after 1979 were significant. Due to agrarian reform unemployment fell from about 32 per cent in 1979 to 20 per cent in 1980, and unionisation has got well underway (Petras 1981).

Agrarian reform programmes have differed in their emphasis, whether it be on land distribution or agricultural production. Often it seems that although there is in theory a clear commitment to land reform, in practice programmes concentrating on higher levels of production are preferred. The first agrarian reform in the region took place in Mexico after the revolution and went through its most radical phase in the 1930s, when much good land was turned over to the *ejidos*, the public sector. During the 1940s the government, having decided that sufficient redistribution of land had taken place, began to concentrate on increasing agricultural productivity. Today the major part of the public sector operates below subsistence level, while production is concentrated in a small number of large, efficient private enterprises.

In other countries, despite the rhetoric of the Alliance for Progress that stressed social justice, the officials, bureaucrats, agronomists and administrators frequently implemented and supported programmes which demonstrated that they could bring about higher levels of production. It is the large farm, with its abundant land and capital and modern machinery, that has invariably been perceived as capable of greater yields. Indeed the commitment to size for greater productivity is demonstrated by the socialist countries, which own large modern farms but replace the private owner with the state.

This assumption that big means better is questionable. Many commentators have noted the fact that the small plot of land may produce more per acre than the large, because of necessity it is cultivated more intensively (see *CEPAL Review* 1982, for example). The Ecuadorian peasantry seem to be particularly productive. A study of the Ecuadorian censuses for 1954 and 1974 estimated that production on the smaller units had grown by an average of 2.7 per cent per year during the period, but on the larger units, growth was at the rate of 1.2 per cent. The Bolivian peasantry has also demonstrated impressive rates of output. Between 1950 and the mid 1970s in the Bolivian Andes, the production of cold- and temperate-climate crops grown by the peasantry expanded at an average annual rate of 4.4 per cent, particularly high levels being recorded in the 1950s after agrarian reform (Ortega 1982).

FOOD POLICIES

Agrarian policies have come to be focused less on land tenure arrangements and more on the growth and distribution of food. If there are problems of malnutrition and hunger, these can be tackled at source by attempts to produce more food. Greater food production also means improved incomes for farmers. To understand government policies concerning food, we need to look at nutritional levels and the quantity and nature of food production.

Statistics for calorie consumption in the region show that despite considerable variation between different areas nearly all the countries raised their calorie requirement satisfaction rates during the 1970s, the greatest progress being made by the countries with initially the lowest levels. According to these rates the best-fed people of the region live in Argentina, Costa Rica, Cuba, Mexico, Paraguay and Uruguay, while Bolivia and Haiti still maintain low levels of calorie intake. Dietary changes also reveal improvements in food consumption patterns, since there has been an increase in the consumption of oils, chicken meat, eggs and milk, which in the past have been beyond the reach of many of the poor. Sugar contributes more to calorie intake than in the past, mainly due to the increased consumption of processed beverages and foods, which may bode ill for the population's teeth, but also reveals increased purchasing power. The declining contribution made by cereals and pulses to calorie intake affects the low income groups (López Cordovez 1982). It could be interpreted as a shift towards other foodstuffs or as a worrying decline in the basic calorie consumption of the poor.

Latin America's nutritional levels are higher than in many other parts of the Third World, and yet it is estimated that 15 per cent of the region's children suffer from medium to high-level malnutrition, which means, given the differences within the region, high levels in some areas (López

Cordovez 1982). One of the major factors inhibiting greater improvements in consumption levels is inflation, to which food prices become very vulnerable. Because of inflation, real food prices in the great majority of Latin American countries were higher at the end of the 1970s than the beginning. It is estimated that 60 per cent of the rural population of Latin America lives in conditions of poverty (ECLAC/FAO 1985).

When it comes to the level of food production, it needs to be pointed out that food production rates for the region are not unimpressive. FAO statistics show that in the period from 1971–80, Latin America had the highest growth rate (3.9 per cent) in food production of the various world regions (Figueroa 1985). Increased yields were made possible by mechanisation and the expansion of the cultivated area. With the recession of the 1980s, however, agricultural production grew by only 1.6 per cent from 1981 to 1984 (ECLAC/FAO 1985). Part of the problem lies in the fact that much of this produce is exported and therefore land which could have been utilised to feed a domestic market is supplying an overseas population. In the early 1980s, 80 per cent of agricultural exports were made up of the following items, in order of importance: coffee, sugar, soya beans, oil-seed meal and oil-cake, cotton, cocoa, bananas, beef and live cattle, maize and wheat (López Cordovez 1982). Governments argue that they need these exports to obtain foreign currency to finance the region's debts.

Agribusiness, which has been moving into the luxury end of the market, with the growth of carnations in Colombia and strawberries in Mexico for sale in the USA, has contributed to the high level of exports. Agribusiness has also found a profitable outlet in the production of animal feed. Much of the Colombian countryside has been turned over to the growing of soya and sorghum to feed chickens, thus depriving the Colombian peasants in the area of a food supply and contributing to the increasing deficit of protein in Colombia. Similarly, most of the Peruvian anchovy catch is turned into fishmeal to feed chickens.

At the same time, the volume of agricultural products being imported to the region is increasing. This increase was at an annual rate of 10 per cent between 1975 and 1980, due to increased purchases of wheat, maize, sorghum, vegetable oils, dairy products, beans and sugar, all basic foodstuffs. A third of these imports came from the region itself, but over 60 per cent came from the industrialised countries (López Cordovez 1982). Wheat has come to be a significant import, which reflects the increasing consumption of bread as a staple, rather than maize-based food products such as *tortillas*. Armstrong and McGee make a more explicit connection between the growth of agribusiness and commercial farming on the one hand and increase in agricultural imports on the other. Between 1960 and 1978, 'Food imports rose in a number of countries where agribusiness and local commercial farmers have been most active in modernising the sector (Argentina, Colombia, Guatemala and Mexico)' (Armstrong and McGee 1985: 77).

What is required in Latin America is a policy that will stimulate food production for the raising of income levels of the rural poor. Given a situation where land is being used to provide agricultural produce for export, one solution to the problem of food shortages would be to make the land that supplies the domestic market more productive. Scientific breakthroughs made this seem possible and gave rise to the Green Revolution.

This international campaign, aimed at increasing the productivity of land by means of the introduction of science-based technology had as its goals freedom from hunger for the populations of LDCs and also a freedom from food dependence. Latin America is not alone in the world in being an agricultural producer and at the same time dependent on imported foodstuffs. The aims were to be implemented through accelerated increases in food production, which were made possible by the discovery of high-yielding grain at two international research centres in Mexico and the Philippines. Between 1953 and 1970 the high-yielding seeds which were introduced in a number of LDCs, such as Mexico, Colombia, India, the Philippines, Sri Lanka and Indonesia, often led to more abundant harvests of rice or wheat. In Mexico, entrepreneurs in wheat production achieved quite a high level of profitability.

In order to get the best results from the new seeds, technology in the form of machinery, chemical products and fertilisers was necessary, and therefore capital was required for their purchase. The new cereals were capable of doubling or trebling the produce of a cultivated hectare, but this could only be done by investing in a package of changes. Therefore only those who could afford this, either on the basis of capital or loans, and those who could master the changes and new technology were able to benefit. In Asia, the miracle rice was wiped out in many areas because it had not been treated with the necessary chemicals.

The Green Revolution in Latin America has led to a Kulak path of development. Some have done well, but they are those who have more land, access to capital, higher levels of education and who are generally more socially advantaged. The Green Revolution encouraged the incorporation of the peasant community into the wider society because it operated through the market. Peasants were enticed away from subsistence agriculture because of the potential profits and, in order to buy the necessary equipment and fertilisers and to sell the final produce, they had to work through the market. In Mexico, the private sector was able to take advantage of the new seeds, but the peasants on *ejido* land, often unable to get credit and farming for the first three years without fertilisers, were not so productive. From 1953–65, the period when the Green Revolution was underway, the wheat yields of the *ejido* sector lagged behind those of the private sector (Pearse 1980).

The technology that was needed to cultivate the high-yield grains successfully often had to be imported, which led to the suggestion that food dependency was being exchanged for technological dependency. Moreover, the necessity for greater capital and technological inputs and the move away from a subsistence agriculture involved the peasant in greater risks. In Colombia, many of the peasants were persuaded to give up their coffee and cocoa trees which, though not highly productive, could be relied on for a harvest, and instead to take up seasonal crops such as corn, soybeans and tomatoes. Cultivation of the new crops required tractors, fertilisers and pesticides, and plants such as tomatoes need constant watering and spraying. This implies capital investment and, although credit was available for the Colombian peasants, interest rates were high. Rubbo cites the example of a farmer who borrowed the equivalent of $800 to plant tomatoes. A month before harvesting, the plants were washed away by some unusually heavy rains, leaving the farmer with a loan and interest that he was unable to pay (Rubbo 1975).

The new seeds were a scientific success but in their introduction to the LDCs, too little attention was paid to the social and political context. The programmes were implemented by governments through existing political channels, thus allowing dominant groups to take advantage of the situation. Many countries who received aid to introduce the Green Revolution had repressive regimes whose policies had in the first place brought about the poverty.

The Green Revolution did bring about an increase in the production of food, which is needed for both urban and rural populations. Those who benefited from improved earnings, however, were not the rural poor, but the farmers who were able to take advantage of the opportunities the Green Revolution had to offer. The main problem was that the sociopolitical implications of scientific and technological change were not fully considered.

A growing urban demand for food has tended to stimulate production in the larger and more commercialised enterprises, but this is not of necessity, nor is it always the case. The potential for developing a peasant economy can be seen in the Ecuadorian situation. In the 1970s Ecuador enjoyed an oil boom, which for a few years gave a boost to the economy as a whole, though the gains were felt mostly in the urban sector. This led to an increase in urban employment opportunities and the expanding workforce gave rise to a greater demand for food. Rural–urban migrants were able to take advantage of the former, often sending financial assistance back to their relatives in the countryside and so contributing to increased income levels there. Urban consumer demand was met by an expansion in food production by the farmers in the Sierra, who, although still the poorest members of Ecuadorian society, had benefited from agrarian reform. The consequence was a real, though small, rise in the income of this peasantry (Commander and Peek 1986).

The circumstances of this case are unusual because of the oil boom, and the following economic crisis of the 1980s (which has now eroded the income gains of those small farmers) has demonstrated the inability of the economy to sustain the development that had taken place. The fact that economic growth had been externally generated meant that when the outside impetus (rising world oil prices) receded, there was no continuing growth. Nevertheless, the Ecuadorian case shows that the peasantry can respond to and benefit from the food demands of the urban population.

With the problem of hunger still in evidence in the developing countries, some Third World governments have turned to other strategies to make the best use of resources and organise distribution, the onus being on distribution as much as production. Mexico, as the world's fifth largest oil producer in 1980, was able to take advantage of its large revenues to finance a new development strategy known as SAM, the Mexican Food System. The aim was to use the wealth from one natural resource to fund a system that would return Mexico to self-sufficiency in basic foods. The policies had three main objectives: to increase home production of strategically important food crops (maize, beans, rice and sugar); to organise the distribution of food to the benefit of the rural and urban poor; and to improve the nutrition of particularly vulnerable groups (Redclift 1984b). Peasants were encouraged to cultivate basic food crops such as beans and corn by the raising of their prices and by assistance in the form of free, improved seeds, cheap fertiliser and credit facilities being made available. The government organisation concerned with food distribution was improved and expanded, so that its subsidised outlets could be found in many poor areas, both urban and rural. The nutritional elements were especially innovative. At the time about 35 million Mexicans, over half the country's population, failed to reach per capita daily food intakes of 2750 calories and 80 grams of protein, the most vulnerable of these being women and children. To attack this problem, a Recommended Basic Food Basket was introduced, which meant the subsidising of essential foodstuffs.

Despite what would seem admirable objectives and a sound financial backing, SAM was short-lived. One of the problems was political, for a change in government, accompanied by the repercussions throughout national- and local-level bureaucracy that this entails, occurred in 1982 and altered the commitment of organisations to SAM. Even before the change, many of the policies, by the time they had been through the rather over-bureaucratic structures, became distorted and did not benefit the peasants to the extent that they should have done. In common with many other Latin American countries, it was the weakness of technical and administrative institutions in charge of peasant agriculture that was partly to blame for the failure of a rural development policy. The peasants themselves were, at times, reluctant to trust government officials and resisted interference from the state. SAM did nothing to reduce the investment of American-owned companies in Mexican food production which dominate the fertile irrigated areas (Redclift 1984b). The national economic crisis led to the removal of many of the subsidies on foodstuffs. The National Foodstuffs Commission, replacing SAM, has been set up to implement a government programme that is based on four main strategies: (1) price supports for basic food production; (2) special incentives to producers of priority foodstuffs; (3) the phasing out of subsidies for non-priority foodstuffs; and (4) a plan to tie retail prices of basic foods to increases in the national minimum wage.

National level plans to bring change to the countryside have varied across the region and changed in emphasis. Initially, with the agrarian reform programmes, the stress was on land tenure arrangements, but in many areas the focus then moved to the production of food, and some governments are now concerned with distribution and access to food.

PEASANT MOBILISATION

It has been suggested that agrarian reform is a reformist strategy for undermining the revolutionary potential of the peasantry by providing rural populations with sufficient reforms to prevent any outright opposition to the system. The peasant, who receives a plot of land through redistribution becomes a member of the petty bourgeoisie, identifying his interests with those of the farmer. This view is well expressed by de Janvry and Ground when they say 'the primary role of the reform sector is political. Its function is to stimulate the development of a conservative agrarian petty bourgeoisie and thus reduce the threat of social instability in the countryside' (de Janvry and Ground 1978: 106).

Others have disagreed with this, but from different perspectives. Petras and LaPorte suggest that agrarian reform brings about revolutionary attitudes through the frustration felt by those who have failed to benefit (Petras and LaPorte 1971). Other studies have shown that profiting from agrarian reform can be combined with a more radical political commitment. Bossert found that the peasants in Chile's central valley with the highest levels of political consciousness were precisely those who benefited from agrarian reform. He concludes that agrarian reform can, therefore, play a part in raising the revolutionary potential of the peasantry (Bossert 1980).

The debate over the radical or conservative nature of the peasantry is a long-lasting one which still continues with some vigour. 'The peasants are conservative' view combines early anthropologists and Marxists in its supporters. The anthropologists saw the rural smallholder as the bearer of tradition and wedded to the same way of life as the generations before him. For Marx, the peasant represented a conservative force because of his structural position in society, which separates him from those who might share his class interests. Low levels of class consciousness are exacerbated by clientelism, which links the peasant through personal relationships to someone in another class.

In some cases where peasant movements have taken place they are said to be clientelist-based. Galjart argues that, despite the successes of the Brazilian peasant leagues and rural syndicates, these were not class movements, but 'followings' of an innovative patron. He shows that the organisers of those leagues were not local peasants, but outsiders, often Catholic priests or members of the Communist party. Their achievements, such as wage rises and the setting of minimum wage levels, were not won through class struggle, but received as favours from local government in the expectation of their political support in return. In some instances, the followings of different leaders competed with each other for different favours. In this way, they operated in a patron–client manner rather than as a united class opposing another and winning reforms through class struggle (Galjart 1964).

Contrasting views of the peasantry point to their radical participation in the Mexican and Cuban revolutions, peasant movements such as that of La Convencion in Peru and the high levels of political consciousness of some sectors of the Chilean peasantry. These views stress different factors that encourage peasant mobilisation.

The now famous middle peasant thesis suggests that certain sectors of the peasantry, precisely those which the anthropologists and Marxists labelled as conservative, have the potential for political mobilisation (Wolf 1969). According to this thesis, the wealthy peasant has no reason to rebel and the poor landless peasant is limited in his actions by his dependence on a landowner. Any sign of political activity would cost him his livelihood. The middle peasant, the small freeholder, has the tactical leverage of independence. He is in a better position than others to initiate confrontation and he has some resources at his disposal. Wolf suggests that it is this sector that acts as the catalyst for rebellion, and he shows how, in six major revolutions in the world, it is the middle peasantry who have formed the pivotal group for peasant action. In Mexico, the peasants from Morelos, led by Zapata, were middle peasants fighting to prevent the loss of their land. In Cuba, although Castro did not receive the support of the *colonos* as a class before 1959, half his fighting force was made up of peasants from the Sierra. These Wolf calls middle peasants because, although they were squatting illegally on the land, in practice they owned and worked their small plots.

The debate over the middle peasantry can only be resolved by referring to concrete situations. Kay suggests that the action of a particular peasant group varies according to the nature of the dominant mode of production. 'The middle peasantry may be a revolutionary force during the transition from feudalism to capitalism, a conservative force in capitalism, a fascist force under the threat of socialism (or capitalism in crisis), and a reactionary force under socialism' (Kay 1978: 119). The mode of production is, as always, important and, as Kay suggests, the transition from feudalism to capitalism may well stimulate peasant action. Wolf shows how the areas in Mexico where there was considerable peasant agitation during the revolution, were those where farms were modernising and becoming more capitalist. In Morelos, the landowners, compelled by increased competition and encouraged by improved communications, were expanding their sugar estates and, in so doing, encroaching on peasant lands. Capitalist agriculture was also expanding in the cotton areas of Durango and sugar plantations of Sinaloa, which were other areas of peasant revolt. In the south of the country, in places like Chiapas, where most peasants were landless and there was little modernisation, the countryside was relatively quiet (Wolf 1969).

It can be argued that the peasantry in LDCs today have revolutionary potential because their situation is so similar to that of the urban proletariat in Marx's day. Marx's argument that the proletariat becomes a revolutionary force (because (1) machinery reduces differences, thus homogenising the labour force; (2) the nature of work in a modern factory requires organisation; and (3) the exploitation of the industrial system leads to poverty and alienation) can be applied to rural society in Latin America. Imperialism focused on one or two natural resources, thus creating a homogeneous agricultural proletariat, all doing the same labouring job. The *colono* system and rural proletarianisation, encouraged by agribusiness, led to a factory-like situation. Domination, first by a foreign power and then by an élite, leads to poverty and alienation. This argument puts forward the notion that the peasantry working in these conditions would provide revolutionary potential.

According to Shanin, the peasantry in general have a low level of class consciousness, but in times of crisis this rises (Shanin 1971). This would certainly explain the frequently found apathy towards radical politics but, at the same time, the revolutionary potential when provoked. There is clear evidence that the peasantry have risen in rebellion on historic occasions, but how much likelihood is there of their political mobilisation in the changing context of rural Latin America?

In the Southern-cone countries, the peasantry represent a small sector, and one that is diminishing. But in Central America and the Andean countries, they constitute one of the major groups in society. Reactions here vary according to different situations. Peruvian agrarian reform has had the effect of raising the consciousness of the peasantry. In Ecuador there has been little unionisation, but organisations based on ethnicity have given the indigenous rural population an effective foundation for action. In Guatemala, the Indian population has opposed attempts to seize their land by wealthy landowners and the army. Nicaraguan peasants, having played a vital part in the Sandinista revolution, have participated actively in agrarian reform. Peasant federations in Costa Rica, a country with a reputation for democratic institutions and relative equality, have been protesting at the surprisingly skewed distribution of land by seizing unused land belonging to multinationals.

The peasant movement in El Salvador has developed through the peasants' own ability to organise and form a union in the face of growing landlessness and poverty. Pearce argues that two catalysts, the church and the revolutionary organisations, encouraged the movement to rebel and support the guerrillas in the civil war, but she stresses that these acted as catalysts rather than agents. In the guerrilla areas of El Salvador, the peasants for the first time have the chance to administer their own lives; here they have set up elected councils, along with health and education programmes, and they organise production (Pearce 1986).

CONCLUSION

Changes in the countryside have had different implications for the rural population depending on (1) the nature of the terrain; (2) the extent and type of capitalist penetration; (3) the existence and extent of agrarian reform; (4) access to government assistance; and (5) levels of peasant organisation. The result is a varied picture with interlocking modes of production, but in general these changes have promoted economic and social differentiation. The two aims of increasing productivity and redistribution that have permeated agrarian thinking have often been seen as competing rather than complementary. In response to the multiplicity of factors affecting their lives, the peasantry have shown that they can act positively, making the most of opportunity when it presents itself. In the face of repressive regimes, the peasantry have shown a capacity and willingness to organise and mobilise.

FURTHER READING

Castillo, L., Lehmann, D. (1982) Chile's three agrarian reforms: the inheritors. *Bulletin of Latin American Research*, 1 (2).

de Janvry A. (1981) *The agrarian question and reformism in Latin America.* The Johns Hopkins University Press, Baltimore and London.

Goodman, D., Redclift, M. (1981) *From peasant to proletarian: capitalist development and agrarian transitions.* Basil Blackwell, Oxford.

Pearce, J. (1986) *Promised land: peasant rebellion in Chalatenango El Salvador.* LAB, London.

Pearse, A. (1975) *The Latin American peasant.* Frank Cass, London.

Redclift, M. (1984) *Development and the environmental crisis: red or green alternatives?* Methuen, London.

Stavenhagen, R. (1978) Capitalism and the peasantry in Mexico. *Latin American Perspectives*, Issue 18, 5 (3).

Chapter

7

Urbanisation

One of the major social phenomena in Latin America since the Second World War, and one that has a profound effect on the social transformation of the continent, is the urban explosion. In the 1950s Latin America was predominantly rural; now the majority of its population live in towns and cities. Table 7.1 shows how urban Latin America has become since 1960. The rapidly rising birth rate is only part of the explanation, since the birth rate has not risen as fast as urban populations have increased. Urban population figures have been rising at a much greater rate than rural ones. Roberts, using statistics for 1960, 1970 and projected for 1980, shows how in six of the largest Latin American countries (Argentina, Brazil, Chile, Mexico, Peru and Venezuela) the urban population has grown at a faster rate than has the total population (Roberts 1978). This suggests that the expansion is largely due to rural–urban migration, which is estimated to account for 40–70 per cent of urban growth in the post-war period up to 1970 (Morse 1971), and about 50 per cent since then. These statistics are well supported by local studies of rural areas demonstrating out-migration and urban studies showing an influx of peasants. Though Latin America is more urbanised than most other parts of the Third World, the process of urbanisation is very much a feature of LDCs in general. 'In the space of one generation, two hundred million people have moved from the countryside to the cities of Asia, Africa, and Latin America' (Hellman 1986: 216) From 1950 to 1980 in Latin America, 27 million people migrated from rural to urban areas (Hurtado 1986).

This rapid growth of urban areas reflects a process of concentration.

Table 7.1 Urban and rural population, by country, 1960 and 1990 (thousands)

Country	1960			1990			Average annual urban growth rate		
	Urban	Rural	% Urban	Urban	Rural	% Urban	1961–70	1971–80	1981–90
Argentina	15 176	5 440	73.6	27 887	4 435	86.3	2.2	2.2	1.8
Bolivia	1 346	2 082	39.3	3 748	3 566	51.2	2.7	3.4	4.3
Brazil	32 627	39 967	44.9	112 643	37 725	74.9	5.1	4.1	3.4
Chile	5 217	2 473	67.8	11 314	1 859	85.9	3.3	2.3	2.3
Colombia	7 682	8 257	48.2	23 079	9 899	70.0	4.7	3.5	3.0
Costa Rica	452	784	36.6	1 420	1 595	47.1	4.3	3.7	3.7
Dominican Republic	977	2 254	30.2	4 329	2 841	60.4	6.2	4.9	4.2
Ecuador	1 519	2 894	34.4	5 934	4 653	56.0	4.6	4.8	4.5
El Salvador	985	1 585	38.3	2 332	2 920	44.4	3.7	2.9	2.2
Guatemala	1 286	2 678	32.4	3 628	5 569	39.4	3.8	3.3	3.4
Haiti	573	3 102	15.6	1 840	4 673	28.3	4.6	3.6	3.8
Honduras	440	1 495	22.7	2 245	2 893	43.7	5.6	5.6	5.5
Mexico	19 297	18 723	50.8	64 304	24 294	72.6	4.9	4.1	3.2
Nicaragua	591	902	39.6	2 313	1 558	59.8	5.0	4.4	4.6
Panama	455	650	41.2	1 292	1 126	53.4	4.5	3.2	2.9
Paraguay	631	1 142	35.6	2 030	2 247	47.5	3.3	4.2	4.5
Peru	4 597	5 334	46.3	15 132	6 418	70.2	5.1	3.9	3.1
Uruguay	2 034	504	80.1	2 644	450	85.5	1.3	0.6	0.8
Venezuela	4 996	2 506	66.6	17 859	1 876	90.5	4.4	5.0	3.6
Latin America	**102 229**	**105 407**	**49.2**	**308 838**	**123 097**	**71.5**	**4.3**	**3.8**	**3.2**

Source: Derived from Inter-American Development Bank Report on Social and Economic Progress in Latin America, 1991: 272.

Capital, very often originating from abroad, is being concentrated in industrial centres, not just towns but big cities where it can prosper. People follow in search of work, adding population concentration to capital accumulation. Armstrong and McGee see this as contributing to the growth of countries with more industrialised centres and the decline of the smaller republics, thus intensifying regional disparities (Armstrong and McGee 1985).

Industrialisation has clearly led to economic growth. Expansion in output was fuelled by growing external demand and generally expansive domestic economic policies. Although consumption grew faster than output, leading to a rapid expansion of imports, the generally improving terms of trade rendered the trade gap small. The very impressive rates of growth reached, especially in Brazil and Ecuador, can be seen in Table 7.2. In the early 1980s, however, these rates were greatly influenced by the economic crisis prompted by the debt problems, with negative rates being recorded for most countries. By the 1990s neo-liberal policies through privatisation were encouraging the return of private capital, particularly foreign to urban areas.

Table 7.2 Growth of GDP and GDP per capita (percentages)

Country	GDP: average annual growth rate (percentage)			GDP per capita: average annual growth rate (percentage)		
	1961–70	1971–80	1981–90	1961–70	1971–80	1981–90
Argentina	4.4	2.5	–1.9	2.9	0.8	–3.2
Bolivia	4.7	4.0	0.1	2.3	1.4	–2.6
Brazil	5.4	8.6	1.3	2.5	6.1	–0.8
Chile	4.2	2.6	2.7	2.0	1.1	1.0
Colombia	5.2	5.5	3.5	2.1	3.1	1.5
Costa Rica	6.0	5.4	2.3	2.5	2.6	–0.5
Dominican Republic	5.1	6.9	1.7	1.8	4.3	–0.6
Ecuador	5.2	9.0	1.7	1.9	5.9	–0.9
El Salvador	5.8	2.6	–0.4	2.3	0.2	–1.9
Guatemala	5.5	5.7	0.9	2.6	2.8	–1.9
Haiti	0.8	4.7	–1.0	–1.3	3.0	–2.9
Honduras	5.3	5.7	2.0	2.2	2.3	–1.4
Mexico	7.0	6.7	1.5	3.6	3.6	–0.8
Nicaragua	7.1	–0.1	–2.4	3.7	–3.1	–5.6
Panama	8.1	5.1	0.5	4.9	2.3	–1.6
Paraguay	4.6	8.7	3.1	1.7	5.5	0.0
Peru	5.5	3.6	–1.0	2.6	0.8	–3.1
Uruguay	1.6	3.0	0.1	0.6	2.6	–0.5
Venezuela	6.3	4.3	0.4	2.7	0.7	–2.3
Latin America	**5.4**	**5.9**	**0.9**	**2.5**	**3.3**	**–1.2**

Source: Derived from Inter-American Development Bank, Report on the social and economic progress of Latin America 1991: 273.

One of the main features of Latin American urbanisation prior to the 1980s, according to Portes, was accelerating primacy (Portes 1989). The focus of urban growth would have been a major, usually a capital, city. Portes, however, shows that although major cities, without exception, continued to grow in the 1980s, their *relative* expansion decelerated. Only two of the countries in the sample providing the data did not experience a decline in primacy. This can be explained by the shift towards export-orientated policies in the 1970s and 1980s which have encouraged the development of industries such as commercial agriculture, forestry, mining and product-assembly that are not located in large cities (Portes 1989).

RURAL–URBAN MIGRATION

Explanations for migration can be offered at different levels. Looking at Latin America as a region, there has been a considerable degree of consensus which fits very broadly with a dependency perspective. The copious quantity of literature written on the subject can be best summed up in the words of Alejandro Portes:

> In Latin America, the process of import-substitution industrialisation has been taken over by subsidiaries of multinational companies that displaced not only domestic producers but workers because these corporations' superior technology, when applied to agriculture, displaced labour from the countryside also. Idle rural labourers headed for the one or two national centres where opportunities for industrial employment existed, only to be confronted with difficult conditions imposed by foreign-led industrialisation (Mangin 1967; Nelson 1969; Leeds 1969; Cornelius 1971). The similarity of these conditions, which were repeated with monotonous regularity from one major Latin American City to another, reinforced the view that the central factor shaping the urbanization process in the region did not consist of idiosyncratic domestic variables but derived from common subordination to external constraints.
>
> (Portes 1989: 8)

Taking an actor orientation it is necessary to examine the motivation of the migrants. These are often described in terms of the push–pull factors, the 'push' being those that drive migrants from their villages, while the 'pull' attract them to the city. 'Push' factors refer to difficulties of earning a living that were described in the previous chapter. It follows that one of the major pulls of urban life is the opportunity of employment, though this is often not borne out in reality. The city also holds out the hope of a better education for the migrants' children, urban facilities, consumer goods and a variety of cheap diversions. The emphasis is very much on the hope of betterment and opportunity; the town offers a fluid situation, where the chances of improvement may arise, unlike the countryside, where opportunities are far more limited. Though conditions are poor in the towns, poverty is not as great as in rural areas.

153

Who are the migrants? With regard to age, they are mostly young adults, who have fewer ties and are therefore more mobile, and who are also more likely to be attracted by the 'pull' factors. The highest rates are to be found among the population between the ages of fifteen and thirty (Ortega 1982). A similar situation exists with African migration where, especially with the early stages of migration, out-migrants are mostly young adults (Roberts 1978).

In virtually all migration statistics, females outnumber males (Gilbert 1974; Bossen 1984; Simmons, Diaz-Briquets and Laguian 1977). It is true that females predominate in most Latin American populations, but the difference between the numbers of males and females in national populations is much smaller than the difference between them in migrant populations. As agricultural work is considered a male occupation, there is little employment available in rural areas for women. In one rural Argentine study, the contribution of the sons to the family productive unit was in the form of agricultural tasks, but the contribution of the daughters took the form of monetary contributions, earned through migration (Miró and Rodríguez 1982). A Chilean study reveals that at the peak migration ages, from fifteen to twenty-nine, there were sixty-two males for every hundred females. For migrants aged between thirty and forty-nine, the ratio was seventy-five males for every hundred females (Simmons, Diaz-Briquets and Laguian 1977). This contrasts with Africa and Asia, where migration is predominantly male (ISS–PREALC 1983). There, men frequently migrate to towns, leaving their family behind to be visited at weekends or to join them in the towns at a later date. Women play a larger part in agricultural activities in African and Asian countries than they do in Latin America.

It is difficult to generalise about the educational level and skills of migrants, who are such a numerous and varied section of the population. Evidence from Peru, Chile and Mexico demonstrates that it is the more educated and skilled who migrate (Roberts 1978). The census data on immigration to Lima since 1940 show that, with the exception of Ancash, migrants come from the departments with the highest literacy rates (Roberts 1978). Roberts goes on to link educational achievement with the level of economic development in the region of out-migration. He proposes a time perspective in the analysis of skill level and the pattern of movement. In the early stages of migration, the peasants come from the richer and more developed rural areas and so frequently have more skills to offer when they reach the town, than those in the less developed areas. With time, migrants begin to come from poorer parts. In the case of Peru, initially the departments that contributed the most migrants to Lima were Ancash, Junín, Ica, La Libertad and Arequipa. All these departments had experienced economic development by large-scale enterprises and often for the export economy. With the exception of Ancash, they also had the highest proportion of people living in urban places. Very few migrants came from poorer regions such as Puno in the south. Latterly, Lima has been receiving migrants from the poorest and remotest departments. Roberts reports a similar trend in Mexico and Brazil, demonstrating that as capitalism penetrates further into peripheral regions, so these areas contribute more to the migratory process.

The migration process may involve more than one stage. Stage migration occurs when a peasant moves to a provincial town for some time and then on to the city. It has been suggested that stage migration is more likely to take place in countries where there is a significant network of small towns and villages such as Chile, Brazil and Colombia. In countries such as Bolivia, where the non-metropolitan population is farm-based, migrants to the capital city are more likely to be rural (Simmons, Diaz-Briquets and Laguian 1977).

PROBLEMS OF SHELTER

The impact of migrants has varied from country to country, but two areas that stand out as being profoundly affected throughout the continent are housing and employment. Initially it was the basic problem of shelter that attracted social scientists, with the result that the main concern was for the home environment of the poor. The rapidly increasing urban population has placed an impossible strain on the provision of housing. The housing deficit in Mexico is about four million units, which means that about 30 per cent of the population is in need of housing (Hurtado 1986). Since the standard housing of towns and cities has been unable to provide shelter, urban populations have been compelled to take the situation into their own hands and put up their own homes.

Thus squatter settlements have sprung up in great abundance, generally in undesirable and inaccessible places, which are unsuitable for standard building. The settlements are often found on steep slopes, such as those in Rio de Janeiro, which cover the hills behind the luxury tourist attractions of Ipanima and Copacabana beaches, or over muddy and polluted swamps on stilts, as in Ecuador. In Guayaquil, 60 per cent of the population live above these swamps and some homes are a forty-minute walk along rickety boardwalks to dry land (Green 1991: 56). Often referred to as shanty towns, these makeshift settlements now house one-third to one-half of the population of many cities. The self-built homes of Mexico City have increased from 14 per cent of the population and 23 per cent of the built-up area in 1952 to more than half the total population and built-up area today (Hurtado 1986).

Each country has a different name for them, often reflecting the local attitudes, such as the graphic *villas miserias* in Argentina, highlighting the poverty of the communities. On the other hand, Peruvian squatter settlements were named 'young towns' in 1968, in order to emphasise the youth of their inhabitants and to suggest a hopeful future for them. The Chileans use their local word for mushrooms, *callampas*, to imply that they simply spring up overnight.

The squatter settlements originally came into being through illegal land invasions. A group of urban poor, some of whom may be recent migrants, driven by insanitary conditions, high rents and overcrowding of the slum areas, squat on a piece of unused land and hastily erect a temporary shack. Since the invasion is illegal, it is often met with some sort of opposition, but in general the shanties remain, mostly because there is nowhere else for their inhabitants to live. The shacks are constructed out of any material that comes to hand, such as planks, cardboard, canvas, thatch, mud, corrugated iron and often rubbish, but these are temporary and often improved upon at a later stage. At first, the settlement consists only of shacks; there are no roads, no water, no street-lighting, though urban facilities may, in time, be extended to these areas. Once in existence, the settlement attracts new migrants, thus expanding quickly. The shanty town of Cuevas in Lima, founded by a group of 500 in 1960, had a population of 12 000 by 1970 (Turner 1970).

In urban areas, the social inequalities of Latin American society are reflected in the increasing spatial polarisation of social classes which has been a feature of urbanisation in the region (Portes 1989). This is not new – there has been a physical separation of classes since the colonial period – but the recent process of urbanisation has exacerbated this situation. The expansion of irregular shelters in peripheral parts of the city has excluded any but the very poor from these areas, while the élite and middle classes have moved away from the urban core to increasingly more exclusive and segregated neighbourhoods. New migrants flock to kith and kin, reinforcing the class nature of settlements. Mexico City has a history of social segregation which has been reinforced by recent demographic trends. Social and spatial inequality in the city 'are an outcome of economic and political processes mediated through a range of sectoral activities: planning, the provision of land, housing and health-care systems, through its transportation, structure and so on' (Ward 1990: 180). In some cases these unintended consequences of social action have been given a helping hand by government policies.

The Chilean authorities, persuaded by free-market ideology that land was not really a scarce resource, removed limits on the operation of the land market. This meant that land ideal for suburban development became available and so could be developed by the rich. Luxurious commercial and recreational facilities emerged in wealthy residential districts, thus consolidating the class character of the area. At the same time rising land prices, due to speculation, forced all but the wealthy out of the market.

The supply of public housing diminished as neo-liberalism deterred the military authorities from buying land or building for the public sector, and tough government policies towards land invasions prevented the growth of new settlements. The housing shortage led to a rise in the number of '*allegados*' or drop-ins, estimated in Santiago in 1987 as being 15 per cent of the metropolitan population (Portes 1989).

The gap between the burgeoning middle-class and élite sectors on the one hand and those in need of shelter on the other was reinforced by government policies of eradicating precarious settlements. Essentially official programmes moved families from squatter settlements in upper- and middle-income neighbourhoods to impoverished sectors, thus separating them from friends and support networks. In 1984 one of these areas, La Pintana, with a population of 148 000 inhabitants, had an unemployment rate estimated at 60 per cent, only two doctors' surgeries and enough schools to provide basic education for half the children in the neighbourhood (Portes 1989). Rural–urban migration along with national- and local-level policies have resulted in a city with low-income groups predominating in the south, a few wealthy areas in a band stretching from the centre to the north-east, and middle-income groups in the north-west.

One of the issues that arises concerning the inhabitants of squatter settlements is their integration into city life, since many of them are migrants and all of them are living in an area which is peripheral, in terms of city services and location. Patterns of adaptation to city life vary across different countries.

The relationship of squatters to urban populations has often been delineated in terms of marginality. The concept implies peripheralisation, an existence on the borders of society, where there is a lack of participation in urban social structure. Marginality may occur on different levels. Cultural marginality refers to a different set of values, life-style and cultural patterns from the urban environment, suggesting that, though they live in a town, migrants may well retain their rural values and traditions. Bonilla argues that the lack of integration of Rio's *favelados* (the inhabitants of the Brazilian shanty towns) into city life is due to their rural values and customs which are different from those of the city-born and which, though modified to some extent by the new situation, still hinder a smooth adaptation (Bonilla 1970). Cultural marginality, which encompasses the culture of poverty concept (Lewis 1966), has been largely discredited because of the implication that traits such as apathy and passivity, which are hallmarks of the culture of poverty, are very strongly imbedded in those who grow up in such a culture and therefore prevent them responding positively if opportunities arise. In many cases, the traits of resignation and low aspiration that Lewis describes are merely responses to the situation in which the poor find themselves, and will change if the situation changes.

Many sociologists have been reluctant to employ the notion of cultural marginality because of the qualitative difference it implies between

157

squatters and their urban counterparts living in the city centre; nevertheless they have used structural marginality to assess the lack of participation in the urban system. Portes argues, from research in Santiago, that marginal man is really one who is in a state of transition (Portes 1970). He found that there was no real cultural difference between Santiago marginals and those in higher stratas, i.e. they shared the same values and aspirations, but the squatters were unable to realise these because of socio-structural features. Both urban and marginal men believed that education was the best means to achievement, but the former had fewer opportunities to actually benefit from schooling. The main difference between the integrated and the marginal man is thus not qualitative but rather quantitative, located in the structure of occupational, educational and housing opportunities.

Although the terms marginal and marginality have frequently been used to address the situation of squatters, they have also been strongly criticised. The notion of marginality carries with it a sense of dualism, since it implies being on the boundaries of urban and rural society, but not integrated into either. Marginality can be criticised on the same grounds, that is to say the separation of rural and urban is not useful since many of the problems of rural areas stem from the integration of the two. Marginal implies that squatters are outside the system, rather than part of it, and therefore they are outside the class structure. The research by Portes shows that the difference between the squatters and the urban population is one of degree rather than kind, the implication being that they are disadvantaged within the system, rather than being outside it. Perlman found the notion of marginality to be a myth in the Brazilian *favelas* because the inhabitants, far from being peripheralised, were involved in urban political activities and leisure clubs and fully aware of what the city had to offer (Perlman 1976). Many of the squatter settlements exhibit high levels of social organisation and stability rather than marginal characteristics. Lomnitz has shown how squatters manipulate social networks to establish themselves in the city (Lomnitz 1977).

A rather different approach is taken by Turner, an architect who worked for some time in Lima (Turner 1965). In reaction to the comments to the effect that many shanty towns are full of poverty, misery and crime, he pointed out that these settlements were an improvement on the city centre slums. The Turner thesis stresses the self-help that is so fundamental to squatter settlements. According to Turner, many migrants arrive in the city and, after spending a short time in the wretched conditions of the slums, decide to improve their living standards by moving to a squatter settlement. This represents an improvement in various ways. In terms of space, sunlight and unpolluted air, the squatters are better off than they were in the crowded, unventilated and noisy slum courts. Best of all, there are no rents to pay and no fear of eviction.

Research in Mexico, Colombia and Venezuela, carried out several years after Turner's work, lends support to some aspects of the Turner thesis, but not others (Gilbert and Ward 1985). Turner's view that migrants move into inner-city slums and then out to shanty towns, is not so much the case now since the large number of spontaneous housing settlements and the widespread knowledge of their existence makes this sort of housing immediately available. In Mexico City and Bogotá, for instance, most new migrants find accommodation in rooms in the squatter settlements.

The argument that these spontaneous squatter settlements are a superior form of accommodation to the city slums is borne out by the consolidation that takes place in low-income settlements. In many Latin American countries, governments have realised that the cheapest way of attempting to solve the housing problem is to upgrade existing housing settlements and ease the development of new sites and services, rather than attempt the seemingly impossible task of constructing buildings sufficient to house the urban population. The kind of assistance that governments have offered has usually been through self-help schemes, the giving out of land titles to the inhabitants or offering loans and strengthening community organisations through the training of local leaders. In Peru, their policies culminated in attempts to turn the young towns into vast cooperative centres by bringing industry to the settlements.

Self-help schemes contributed to the consolidation and upgrading of housing in the settlements, so that in countries like Peru, Chile, Brazil and Venezuela some of the housing is of two storeys and houses are constructed of brick and other regular building materials. Consolidation has been accompanied by an increase in renting, a phenomenon which existed for some time in African shanty towns, before it became common in Latin America. Gilbert and Ward found that renters made up 36.4 per cent of their sample of inhabitants of spontaneous housing settlements in Bogotá, 12.7 per cent in Mexico City and 4.5 per cent in Valencia (Gilbert and Ward 1985). Although the proportion is still small in Valencia, in all three areas of their research (Bogotá, Valencia and Mexico City), it is on the increase. Further research on renters and owners showed that the main difference between them was age. Renters were young and at an early stage of the developmental cycle. Many of the owners had once been renters, thus, suggesting a pattern where young migrants, who cannot afford land or property, rent until they have sufficient savings to set up a home of their own. Renters are not 'down-and-outs', but those who do not have the opportunity to own their own home, though they may well do so in later life.

Another reason why the increase in renting is likely to continue is its relationship to land prices and availability. Gilbert and Ward found the highest ownership levels in Valencia where land is cheap, and the highest levels of renting in Bogotá, where land is expensive (Gilbert and Ward

159

1985). Since land prices, which are related to land availability, are, in general, rising, it would seem likely that renting would increase. The Turner thesis has been criticised by Marxists for over-emphasising the differences between slum-dwellers and squatters. They argue that the differential in living standards between the two groups is minimal and to see this as social mobility is to detract from the important point they are all part of the large urban poor whose poverty is due to the wider social formation and, in particular, the capitalist mode of production. Self-help schemes relieve the state of the immediate necessity of providing housing and offer a cheap source of shelter, thus making it possible for the poor to survive on very low incomes (Burgess 1978).

The criticisms of both marginality and the Turner thesis have suggested that the problems of spontaneous housing cannot be fully understood without some reference to the wider society. The initial founding of a squatter settlement is, in itself, an illegal act and, therefore, a challenge to authority. Latin American governments have responded in different ways, some taking a sympathetic approach through self-help schemes, as in the Peruvian case, but even these may resort to repression. The Peruvian military reacted with force during the Pamplona invasion in May 1971, when they tried to evict the tens of thousands who had participated in this huge squatter invasion in Lima, with the result that there were several injuries and one death (Collier 1975).

Other governments have viewed the spontaneous housing settlements as wretched areas, which not only house society's drop-outs, but are also breeding grounds for crime and radical activity. In the 1960s and 1970s, the Brazilian government tried for these reasons to eradicate the squatter settlements, especially those in Rio de Janeiro. Rio has the added factor of being an international tourist centre where the *favelas* are so visible on the steep hillsides behind the luxury hotels and beaches. Some of the *favelas* have been knocked down, but there still remains the problem of shelter for their inhabitants. As a result, when shacks disappear, others appear in other parts of the city. The government has attempted some *favela* removal programmes, but these were not very successful, mainly because the new housing schemes were too expensive. The great advantage of spontaneous housing is its cheapness and, until governments can provide adequate housing with the same attraction they are not likely to be successful in their eradication programmes. For the government-sponsored Cidade-Alta housing project in Rio de Janeiro, only families earning three minimum salaries were acceptable, which eliminated 40–69 per cent of the clientele. In some cases where *favelados* had moved to new housing schemes, they had been unable to meet the payments, which are periodically increased due to inflation, and therefore had to return to the *favelas*.

Since the 1980s in Brazil, a rather different school of thought has prevailed, one which says that the *favelas* should be upgraded and consolidated by the provision of basic services, infrastructure and community facilities. This policy has been put into practice by municipal departments of social development, in conjunction with local communities. Under this programme the municipality provides planning, technical and social assistance and financing, while the community provides labour and participates in decision-making and administration of the project.

Self-help schemes provide Latin American governments with a cheap solution for housing the poor, but the contribution they make to the wider socioeconomic structure is more far-reaching. As Gilbert has suggested, 'Self-help keeps the Third World economy functioning' (Gilbert and Gugler 1992: 151). For a start, it allows the poor into the housing market by keeping costs low and not threatening higher-income groups. Where there is consolidation, this offers up opportunities to commercial and industrial companies (e.g. where materials such as glass, bricks, cement, tiles and pipes are supplied in large quantities for construction). Marxist arguments support this view, pointing out that cheap housing reduces the pressure on wages, allowing the labour force to reproduce itself despite low wages and so contributing to capital accumulation in the capitalist sector. On a political level, these schemes help to maintain the status quo, because they satisfy some of the minimum housing needs for the poor and so reduce the pressure for radical reforms. As the satisfaction is just sufficient to prevent people resorting to large-scale political opposition, governments are not pressured into more forceful measures, such as urban reform or progressive taxation.

EMPLOYMENT

International factors have played their part in helping to shape employment opportunities in Latin America. The same forces that stimulated rural–urban migration, namely the displacing of domestic workers by the superior technology of multinational corporations, have created limits to labour-absorption in industrial sectors. In addition to this, since the mid 1970s, the decentralisation and flexibilisation of labour of global post-Fordism (mentioned in Chapter 2) have created openings in increasingly specialised sectors, where forms of employment are short-term, irregular and lacking in social protection.

Post-Fordism has resulted from a number of factors: the decomposition of complex production processes, improved transport, better telecommunications and cheap labour in certain parts of the world. Together these mean that it may be more profitable for a company to site its production processes in a developing country, where labour is cheap and communications systems are effective, than in the home country. Expanding production in the LDCs has frequently taken place in 'free production zones' which are enclaves designed to attract foreign capital by offering a range of incentives, such as exemption from taxes and duties, freedom from foreign exchange controls, the absence of militant unions, provision of factory and office space and a variety of supporting services (Hansen 1981). In this way the host country benefits from the expansion of employment opportunities, while the company reduces its production costs and takes advantage of the incentives. The resulting new division of labour reflects a shift from the traditional role of the LDCs as sources of raw materials and markets for goods produced in the industrialised countries to one where they participate more in world industrial production. Since job opportunities in industry have been limited in this way, how do the millions coming into the city earn a living?

Although a small number do find work in factories as semi-skilled or unskilled labourers, a significant aspect of the Latin American economy is the large size of the service sector. Table 7.3 shows how this sector grew during the 1960s, 1970s and 1980s. As a result, the productive sector is employing a relatively small number of the labour force, while the sector that serves the region absorbs greater numbers.

Table 7.3 Latin American employment characteristics, 1965–91

| Country | Percentage of population of working age | | Proportion of the labour force in: | | | | | |
| | | | Agriculture | | Industry | | Services | |
	1965	1985	1965	1989–91	1965	1989–91	1965	1989–91
Argentina	63	60	18	13	34	34	48	53
Bolivia	53	53	54	47	20	19	26	34
Brazil	53	59	49	28	20	25	31	47
Chile	56	63	27	18	29	30	44	52
Colombia	49	59	45	1	21	31	34	68
Costa Rica	49	59	47	24	19	30	34	46
Dominican Republic	47	53	59	46	14	15	27	39
Ecuador	50	53	55	30	19	24	26	46
El Salvador	50	60	58	10	16	35	26	55
Guatemala	50	53	64	48	15	23	21	29
Haiti	52	51	77	50	7	6	16	44
Honduras	50	50	68	36	12	17	20	47
Mexico	49	54	49	22	22	31	29	47
Nicaragua	48	50	56	46	16	16	28	38
Panama	51	58	46	12	16	21	38	67
Paraguay	49	51	54	48	20	21	26	31
Peru	51	56	49	35	19	12	32	53
Uruguay	63	63	20	15	29	18	51	67
Venezuela	49	56	30	12	24	32	46	56

Source: Derived from *Human Development Report*, 1993: 168 and *Statistical Abstract of Latin America*, 1993, Vol. 30, Part 1: 387.

In Latin America a number of those who do find jobs in the productive sector work in *maquiladoras*, the 'free production zones' of the region. These assembly plants, found particularly in engineering and electronics, have expanded in numbers since the mid-1960s. Mexico offers very specific advantages, for its proximity to the USA means that factories sited in northern Mexico have access to North American facilities such as good transportation networks; at the same time there is an abundance of cheap labour. It follows that assembly plants are to be found predominantly in the border towns, where in some cases North American managers can live in Texas and work in Mexico. In Ciudad Juarez, probably the most important centre, there were five *maquiladora* firms in 1966, but by 1983 the figure had grown to 135 (Carillo and Hernandez 1985).

The *maquiladoras*, however, have been a mixed blessing for Mexico. They do provide a source of employment – 71.5 per cent of industrial employment in Ciudad Juarez is in the *maquiladoras* and this employment is not threatened by local economic fluctuations (Carillo and Hernandez 1985). It is notable that these industries have been less affected than others by the crises of the 1980s. They also make an important contribution to foreign exchange since their earnings are now second only to oil.

On the other hand, as we saw in Chapter 5, the work on offer receives very little pay and invariably involves poor working conditions. In Ciudad Juarez, where about 80 per cent of their employees are women between the ages of fifteen and twenty-five, there is a high turnover rate, often due to eye and back problems resulting from poor lighting and seating arrangements. Most of this young labour-force participate in unskilled work, so there is no educational benefit (as is sometimes suggested) from learning new skills. Few of the managerial and technical staff are Mexican. The lack of restrictions and general 'red tape' which is so attractive to employers has resulted in particularly high levels of pollution that are eventually receiving some attention now that they are threatening US cities to the north.

INFORMAL SECTOR

Despite the lack of opportunity in regular wage-earning work official unemployment figures are not very high (see Table 7.4). This is largely explained by the statistics for 'disguised' unemployment or underemployment, which often reach striking levels (see Table 7.5).

Since Latin American governments offer no real assistance to the unemployed, many of the poor are compelled to earn a living through irregular and haphazard means. The category of underemployment refers to an area of work that includes street selling, shoe-shining, refuse collecting for sale, prostitution and many other activities which offer no security and bring in very meagre earnings. Table 7.5 reveals a slight contraction in this sector during the 1970s, but nevertheless in 1980 it still represented over a third of the economically active population for the region as a whole. The economic crises of the 1980s have contributed to its expansion in this decade.

Table 7.4 Urban unemployment

Country	1975	1980	1985	1989	1990	1991
Argentina	3.7	2.3	6.1	7.6	7.4	6.5
Bolivia	–	7.1	5.8	10.2	9.5	8.1
Brazil	–	6.3	5.3	3.3	4.3	5.0
Chile	15.0	11.8	17.0	7.2	6.5	7.9
Colombia	11.0	9.7	14.1	9.9	10.3	10.3
Costa Rica	–	6.0	6.7	3.7	5.4	5.0
Ecuador	–	5.7	10.4	7.9	–	–
Guatemala	–	2.2	12.0	6.2	6.4	6.5
Honduras	–	8.8	11.7	8.0	7.1	8.4
Mexico	7.2	4.5	4.4	2.9	2.9	2.6
Nicaragua	–	22.4	22.3	–	–	–
Panama	8.6	10.4	15.7	16.3	16.8	15.1
Paraguay	–	3.9	5.1	6.1	6.6	–
Peru	–	7.1	10.1	7.9	8.3	–
Uruguay	–	7.4	13.1	8.6	9.3	9.2
Venezuela	8.3	6.6	14.3	9.7	10.5	10.9

Source: Derived from *Statistical Abstract of Latin America*, 1993, Vol. 30, Part 1: 410.

Portes argues that informal labour does not 'function as an effective counter cyclical mechanism against the contraction of the modern sector. Instead both informality and open unemployment grew (in the 1980s) simultaneously in most countries' (Portes 1989). In the 1970s both open unemployment and informal unemployment declined or remained static through most of the decade, while in the 1980s both rose quite markedly. For Latin America as a whole open unemployment rose from an average of about 6 per cent of the urban economically active population in 1974 to some 14 per cent in 1984. That was the year in which urban unemployment reached its peak in Colombia, Peru, Honduras and Venezuela (Portes 1989). In countries such as Colombia and Uruguay, much of the expansion of informal labour took place among women. As they are often employed as outworkers for clothing manufacturers or in the leather export industry, working at home in sweat-shop conditions, this increased flexibility and reduced costs.

Some social scientists have conceptualised these workers in terms of a reserve army of labour. In this case, they would constitute a pool of labour which can be utilised in boom periods and disregarded in recessions. This view sees the experience of the Third World as being little different from that of the industrialised countries. The large numbers who are unable to find regular work are not seen as a phenomenon characteristic of under-development, but as part of the capitalist division of labour and, therefore, a similar feature to the industrialisation process of European countries, differing in terms of quantity and not quality. For Marx, however, the function of a reserve army of labour is keeping wages low and, to this end, the absolute size of the reserve army need not be very great. The high numbers and continuing growth of this sector in Latin America would suggest that concepts of European analysis are not sufficient.

This sector has also been referred to by some as lumpen proletariat. The implication here is that these workers have rejected the dominant values of society and are made up of criminals, prostitutes, drug-addicts and other drop-outs (Lloyd 1982). This image clearly does not corres-pond with the material presented so far in this chapter.

Others use the term sub- or proto-proletariat to mean that these mem-bers of the urban poor are in the process of becoming a working class, though have not yet achieved that status. It implies a commitment to the urban way of life and urban employment, and the final development of this sector, at some time in the future, into a proletariat proper (Lloyd 1982). The expansion of the sector in the 1980s presents a genuine diffi-culty with this concept.

The terms underemployment and unemployment have a negative ring about them and do little to help our understanding of this type of work. In order to shift the emphasis towards a more positive interpretation, Keith Hart in 1973 introduced the concept of the informal sector (Hart 1973). The division between formal and informal is based on the distinc-tion between wage-earning and self-employment, with the key variable being the rationalisation of work, in other words whether or not labour is recruited on a permanent or regular basis for a fixed reward. The empha-sis on self-employment means that new income-generating activities can be identified. The term informal sector has become widely used.

What are the characteristics of the informal sector? For a start, there is considerable freedom of entry for those involved, unlike the formal sec-tor where barriers exist in the form of qualifications, contracts and other institutional restrictions. Secondly, this sector is highly competitive, there is a complete absence of the monopolies which predominate in the formal sector. Linked to this is the absence of foreign ownership within the informal sector, the entrepreneurs and pedlars are local people. Finally, pricing mechanisms operate in a different way from the formal sector, where they are fixed. They fluctuate according to the personal factors of those involved in the transaction (Davies 1979).

This sector supplies both the poor and the rich and produces both goods and services. Although all these activities lack security, they vary in the extent of their irregularity and in the nature of the work involved. Street sellers of different kinds are an important group, whether they be

165

Table 7.5 Latin America: segmentation of the economically active population and underemployment coverage: 1950 and 1980 (percentages)

		Share in the total EAP						Underemployment coverage
		Non-agricultural			Agricultural			Mining	
		Formal	Informal	Total	Modern	Traditional	Total		
		(1)	(2)	(3)	(4)	(5)	(6)	(7)	(8) = (2) + (5)
Latin America	1950	30.6	13.5	44.1	22.1	32.6	54.7	1.2	46.1
	1980	47.7	19.4	67.1	13.2	18.9	32.1	0.8	38.3
Group A	1950	26.4	12.2	38.6	22.4	38.0	60.4	1.0	50.2
	1980	48.2	18.6	66.8	14.1	18.4	32.5	0.7	37.0
Mexico	1950	21.6	12.9	34.5	20.4	44.0	64.4	1.1	56.9
	1980	39.5	22.0	61.5	19.2	18.4	37.6	0.9	40.4
Panama	1950	34.9	11.8	46.7	6.2	47.0	53.2	0.1	58.8
	1980	51.6	14.8	66.4	11.4	22.0	33.4	0.2	36.8
Costa Rica	1950	29.7	12.3	42.0	37.3	20.4	57.7	0.3	32.7
	1980	54.2	15.3	69.5	20.5	9.8	30.3	0.2	25.1
Venezuela	1950	34.7	16.4	51.1	23.3	22.5	45.8	3.1	38.9
	1980	60.9	18.5	79.4	6.5	12.6	19.1	1.5	31.1
Brazil	1950	28.5	10.7	39.2	22.5	37.6	60.1	0.7	48.3
	1980	51.6	16.5	68.1	12.4	18.9	31.3	0.6	35.4
Colombia	1950	23.9	15.3	39.2	26.2	33.0	59.2	1.6	48.3
	1980	42.6	22.3	64.9	15.8	18.7	34.5	0.6	41.0

Group B	1950	17.1	14.9	32.0	23.2	43.0	66.2	1.8	57.9
	1980	29.1	21.8	50.9	12.0	35.9	47.9	1.2	57.7
Guatemala	1950	16.6	14.0	30.6	20.6	48.7	69.3	0.1	62.7
	1980	23.8	18.9	42.7	19.4	37.8	57.2	0.1	56.7
Ecuador	1950	21.5	11.7	33.2	27.4	39.0	66.4	0.4	50.7
	1980	25.6	28.6	54.2	12.1	33.4	45.5	0.3	62.0
Peru	1950	19.1	16.9	36.0	21.9	39.4	61.3	2.7	56.3
	1980	37.7	19.8	57.5	8.9	31.8	40.7	1.8	51.6
Bolivia	1950	9.1	15.0	24.1	19.0	53.7	72.7	3.2	68.7
	1980	17.9	23.2	41.1	5.2	50.9	56.1	2.8	74.1
El Salvador	1950	18.5	13.7	32.2	32.5	35.0	67.5	0.3	48.7
	1980	28.6	18.9	47.5	22.3	30.1	52.4	0.1	49.0
Group C	1950	54.0	16.6	70.6	20.4	7.6	28.0	1.4	24.2
	1980	61.5	21.4	82.9	9.2	7.0	16.2	0.9	29.4
Argentina	1950	56.8	15.2	72.0	19.9	7.6	27.5	0.5	22.8
	1980	63.5	21.4	84.9	7.8	6.8	14.6	0.5	28.2
Chile	1950	40.8	22.1	62.9	23.1	8.9	32.0	5.1	31.0
	1980	55.5	21.7	77.2	13.2	7.4	20.6	2.2	29.1
Uruguay	1950	63.3	14.5	77.8	17.3	4.7	22.0	0.2	19.2
	1980	63.3	19.0	82.3	9.5	8.0	17.5	0.2	27.0

Source: Garcia and Tokman, *CEPAL Review*, 24: 105.

street pedlars who manage to buy chewing gum, newspapers, sweets, balloons, ornaments, fruit and any number of miscellaneous items wholesale, or the women who make snacks for sale in the streets. A good illustration of the latter can be found in Domitila Barrios de Chungara's account of the daily lives of a Bolivian miner and his family. She relates how she, like many other miners' wives, sells foodstuff in the street in order to make up for what her husband's wage does not cover in necessities. Domitila makes and sells pies and because this requires all the tasks of preparing vegetables, buying the meat, cooking the pies and then selling them, the children help her. Shopping at the company store often means a long queue, so the children line up for the groceries while Domitila sells her pies (Barrios de Chungara and Viezzar 1978).

This sector also includes outworkers for manufacturing and repair operations, who subcontract work to do in their homes or small workshops, such as shoe manufacturers (Peattie 1982); beggars, who are often more organised than they appear (Ruiz-Perez 1979); prostitutes; garbage pickers who can sell various types of waste paper, refuse and bottles; the live-in domestic servants paid low wages and provided with board and lodging, who according to Lloyd are included in the informal sector by default (Lloyd 1982), and others who work on some sort of irregular basis.

Because of the varied nature of this sector, and because of criticism of the term informal sector *per se*, other attempts to clarify and analyse this area have been made. Informal sector implies a dualist interpretation of the urban economy, since it proposes a dichotomy between the formal modern capitalist sector in which big business and multinationals flourish, and the mass of the poor who are unable to benefit from participation in this sector. The implication is that poverty is due to lack of involvement with successful capitalist enterprise and though a minority of the poor do become successful entrepreneurs by building up small businesses from street selling, the vast majority are poor because they are excluded from the modern economy. There is also a problem in delineating this sector since self-employment is considered such a crucial factor. Occupational statistics that deal with categories of employment make no distinction within the self-employed group. As a result, a chewing-gum seller and a successful carpenter who has his own business both fall within the same category.

The important point to focus on is the relationship between the modern economic sector and the informal sector which is clearly not as negative as the formal/informal dichotomy would imply. Many social scientists see a direct link between these sectors similar to that which exists between rural and urban society and between Latin America and the metropolis: i.e. it is a relationship of exploitation. It is not because of the poor's isolation from the modern sector that they remain poor, but because they provide a cheap human resource for that sector.

Moser points out that the majority of small-scale enterprises of the type described in the informal sector fit into the character of petty commodity production (Moser 1978). This allows us to refer to Marx's analysis of the phenomenon. He argues that petty commodity production exists as a subordinate form in all modes of production, but thrives particularly in the transition from feudalism to capitalism. It contributes to the development of capitalism in two ways. Firstly it has its roots in petty production and private ownership of property, and, as such, forms the basis for the development of capitalism through increases in the scale of production. Secondly, it acts as a source of capital and labour which can be indirectly used by other forms of capital that may be operating in other spheres of the economy. It is this second point which is especially relevant to the case of Latin America. The Marxist view is that petty commodity production indirectly contributes to the surplus in the modern economy, which is so necessary for capital accumulation (these points are well developed by MacEwen Scott 1979). The argument is, then, that a greater surplus can be extracted from this sector by not employing its members on a regular wage-earning basis. This is a similar argument to the Burgess critique of Turner, emphasising the way that self-help shelters the poor cheaply, thus making low wages and the resulting capital accumulation more feasible.

Quijano, among others, takes this argument a stage further by utilising the modes of production approach. He contests that petty commodity production is a separate mode of production to the capitalist one, but that it articulates with it to facilitate expanded reproduction of the capitalist mode (Quijano 1974). The relationship between capitalist and non-capitalist modes is one of exploitation, in which the former creams off the surplus from the latter. In this way, it benefits capitalism to maintain non-capitalist modes of production, rather than engulfing them within the rubric of the capitalist mode. In order to fully comprehend the nature of the relationship between the two sectors of the economy it is necessary to look at empirical material.

THE GARBAGE-PICKERS OF CALI

A significant indicator of poverty in any society is the number of people who can be seen poking in rubbish bins looking for something that is still usable, since surviving on society's refuse epitomises ultimate poverty. Birbeck's study of garbage-pickers in Cali, Colombia, suggests that they should not be viewed as vagrants left behind by economic development but as workers who are part of the industrial system (Birbeck 1979). He argues that this activity should be seen, not just as an expression of poverty, but as a cause of it. The garbage-picker may be excluded from real opportunities within the urban economy, but, nevertheless, his world is closely integrated into the modern economy.

The paper industry in Cali has two sources of raw materials – long-fibred softwood pulp is imported from Canada, Scandinavia and Chile to make the best quality paper, while waste paper is recycled for poorer quality products. In Cali, about 25 000 tons of waste paper are collected each year, of which the garbage-pickers contribute 15 000 tons. As is typical in LDCs, the paper company has experienced foreign penetration and is dominated by one giant company. Cartón de Colombia was founded in 1944, largely on the basis of North American capital that is now in the hands of Mobil Oil Company. Although laws were passed in the 1960s reducing the share of US capital in the firm, it still remains the major shareholder. The waste paper necessary for the industry is bought from warehouses which are not part of the Cartón de Colombia company. There is a contractual relationship between firm and warehouse, saying that all waste paper, with a few exceptions, will be sold to the factory at a fixed price. The warehouses are organised in a hierarchical manner with a central one buying from satellite warehouses, which are situated in poor residential areas of Cali, so that they are near the homes of the garbage-pickers.

The garbage pickers are not employed by Cartón de Colombia or the warehouses, nor do they have any sort of contact with them. The price of goods varies for the same item, in fact Birbeck noted as many as ten different prices for the same type of waste paper. The price is more likely to relate to the individual picker and the regularity with which he sells to the warehouse. Buyer and seller strike a bargain with each individual purchase. It is difficult to assess the number of garbage pickers because this fluctuates seasonally and they are not really enumerated in official statistics, but Birbeck estimates that at the time of the study there were between 1200 and 1700 garbage-pickers in Cali. This is not a startlingly high number but, bearing in mind that each probably supports a family, some 5000 to 10 000 may well depend on an income from garbage-picking. The pickers operate in different ways. The largest group work at the municipal garbage dump near the banks of the River Cauca. They collect here each day when the municipal truck appears, and go through the refuse looking for resaleable objects, which are put in a sack to be sorted out later. Others intercept at an earlier stage, working along the routes that the municipal trucks take. In between the time the dustbins are put out and the truck arrives, the pickers have been through the contents of the bins. Others go around residential neighbourhoods with their carts collecting different kinds of scrap from shops and houses.

Due to the irregular method of payment, the amount of income is low and variable. Most pickers earn the equivalent to the government minimum daily wage in a day's work, but they do not get any of the benefits of Social Security payments, minimal though they are, that the employed workers receive, nor do they have the security of a regular wage.

As most of the domestic waste for recycling is collected by garbage-pickers, why do Carton de Colombia not employ the garbage-pickers? Clearly it is cheaper for them to operate in the way they do because they are not obliged to pay regular wages. The income the garbage-picker receives for each item is extremely low and reduced even further by the competition between pickers which is encouraged by the system that is very much a buyer's market. The small regular labour force that is employed by the company is relatively well paid, receiving three times the minimum wage. To take the garbage-pickers on at this rate would increase the company's costs and would also bring about the additional cost of security benefits. The system also gets round problems of fluctuation: when business is slow the company does not have to maintain a regular workforce.

There is another relevant factor, and that is that the labour is kept fractionalised. The small but relatively well-paid labour force is kept separate from the garbage-pickers who are further divided between themselves by intense competition. The labour force realises its relatively privileged position and therefore does not embark on serious attempts of collective bargaining. It is significant that the union that represents the workers of Cartón has never been on strike.

A final factor contributes to the low income of garbage pickers, and this relates to their involvement in the international economy. Of the two materials that are used by the paper company, pulp is always preferable to waste paper and in Cali pulp can always be imported. For waste to be a competitive commodity, it must be distinctly cheaper than the pulp. If garbage-pickers were to become regular employees, the price of waste paper would rise by at least three times and it would not be worthwhile for the company to buy it.

The garbage-pickers appear to work for themselves, but in fact are part of an industrial organisation. Through their relationship with big industry they reduce costs by piecework and division of the labour-force, so contributing to capital accumulation in the modern sector. For the garbage pickers themselves, however, their relationship with major industry simply represents a poverty trap.

POVERTY TRAP OR ENTREPRENEURSHIP?

A related difficulty with the concept of informal sector arises from the range of activities it covers. Bromley points out that there is a great difference between the truly independent street seller and seasonal workers, who work for a regular wage though not on a permanent basis and who would be included in the informal sector category (Bromley and Gerry 1979). He prefers to use the terms casual work to cover this broad area, and defines it as 'any way of making a living which lacks a moderate degree of security of income and employment' (Bromley and Gerry 1979: 5). Since the similarities between informal sector seasonal workers and formal sector factory workers are greater than between some groups within the informal sector, such as street sellers and seasonal workers, Bromley suggests a continuum of categories as a more useful form of analysis. The continuum stretches from stable wage work to true self-employment. It can be divided into four broad and occasionally overlapping categories, which in order of decreasing similarity from stable wage work, are short-term wage work, disguised wage work, dependent work and true self-employment at the polar extreme of the continuum.

Short-term wage work is paid for and contracted for a specific period of time, whether that be a day or a season. This would include workers employed in tourism for a holiday period, shop assistants who are employed for a Christmas or summer season and construction workers who are employed on a contractual basis.

Disguised wage work refers to work where one or more enterprises appropriate part of the product of a person's work without the person officially being an employee of the firm. Many firms in manufacturing and repair operations use outworkers who perform the work in their own homes, being paid for each completed piece. Small clothing shops in Lima frequently operate in this way. Merchants purchase the materials and prepare the parts ready to be put together as garments, which are either collected by or delivered to the outworkers, who sew the pieces together. The garment has to be completed within a certain time, then the outworker is paid a requisite sum. In this case, the firms relinquish responsibility for the instruments of production, which are provided by the workers themselves (MacEwen Scott 1979).

With dependent work, the workers appear to be self-sufficient, but in reality are dependent on a firm or group of firms for some vital part of their work, such as premises, equipment, supplies, or outlet. In this case, a proportion of the product of the worker's labour is appropriated through the payment of rent, repayment of credit or purchases of supplies at prices disadvantageous to them. Many taxi-drivers in Latin America, unable to afford their own vehicle, have to rent one; thus they have to deduct fixed rental and running costs from their profits, which are considerably diminished as a result. In a similar position would be a street trader, who regularly deals with a certain wholesaler and who could only gain credit from him by guaranteeing to purchase his goods. In this way, the street trader becomes dependent on the wholesaler.

The truly self-employed works quite independently, without involving himself/herself in any form of wage work. This category, which has the least security but the greatest freedom of manoeuvre, includes street pedlars (who buy from wholesalers according to availability and cost) and those who make up foodstuffs or cook for sale.

One importance of the continuum is that it is more precise form of categorisation than the simple dichotomy. But what is more important is that it outlines the links between casual work and the formal economy, setting these out in terms of declining significance. In each case the links are exploitative. The seasonal workers and outworkers do not receive the security or the minimal benefits that are available to regular workers. The firm saves on this and, at times, provision of facilities or equipment, thus making it a profitable exercise for the managers. The taxi-driver and the street trader, once tied to debts or rental, cannot accumulate sufficient capital, and so can be further exploited. Even the truly independent are dependent on social and economic conditions, which are favourable to them. They are operating at the bottom end of the market with no capital and no security.

Poverty is compounded by inflation which has reached extreme levels in the region, as demonstrated by consumer price indices. Between the initiation of the Chilean price index in December 1978 and April 1983, the index registers a 157.5 per cent increase for low-income families and a 142.7 per cent increase for higher-income groups. In the twelve-month period from February 1982 to February 1983 alone, consumer prices rose 42.9 per cent (LAWR 1983).

In LDCs in general, the poor are disadvantaged because they have low levels of education and training, and therefore few skills or qualifications to offer. Because of their position in the class structure they do not have access to useful connections which, as we have seen, are so valuable in Latin America. Caught in the poverty trap, they are unable to make the savings necessary for business ventures. Their plight is reinforced by their close proximity to wealth.

For a growing number of social scientists, planners and government authorities, the informal sector is coming to be seen as a source of entrepreneurship rather than a poverty trap. Gwynne suggests that this aspect has been neglected by researchers, who have concentrated on street pedlars and garbage-pickers rather than the small-scale manufacturing sector, which does have the potential for growth (Gwynne 1985). He cites the entrepreneurial success of modular furniture-makers in Caracas. Because high-rise flats became so popular in the city in the 1970s, the demand for modular furniture appropriate for small residences rose. This was answered in part by a number of small-scale entrepreneurs operating in the shanty towns. Although based in the informal sector, they used some modern-sector facilities, such as newspapers, to advertise their work, and because of lower costs were able to compete successfully with large furniture manufacturers. The entrepreneurs in this case were able to take advantage of the low cost of the informal sector and, at the same time, to make use of the modern facilities of the formal sector. This close interlinking of the sectors benefited the entrepreneurs.

Currently, Latin American governments are showing less and less desire to remove or absorb the informal sector. The initiative, innovation and great desire to work which exists within the informal sector is

173

highlighted by certain policy-makers, instead of the criminal or apathetic aspects of yesterday. Hernando de Soto, a Peruvian member of Congress, believes that there is abundant entrepreneurial talent within the informal sector. Like Safa, he talks of the ingenuity and adaptive capabilities of the informal sector workers (Safa 1982). On the one hand de Soto seems to be suggesting that development is only possible if legal institutions are available to all, while on the other he is lauding the entrepreneurial opportunities to be seized by working within the informal sector. Ultimately he is attracted by the ideal free market of the informal sector, a sector which is 'not beset by the legal and bureaucratic problems which hinder the formal sector' (de Soto 1989).

De Soto cites a 1983 study by the Institute for Liberty and Democracy in Peru who set up a fictitious garment factory and went through the bureaucratic procedure of doing it all legally – a process which took 289 days and cost $1231, which is the equivalent of thirty-two times the minimum wage. His comment concerning the above study states that 'were it not for the costs of remaining formal, the firm's profit and therefore savings and potential investment capital would be quadrupled' (de Soto 1989: 148). In his conclusion he states that:

> competitive business people, whether formal or informal . . . provide a sounder basis for development than sceptical bureaucracies . . . . In Peru, informality has turned a large number of people into entrepreneurs, into people who have known how to seize opportunities by managing informal resources, including their own labour, relatively efficiently. This is the foundation for development, for wealth is simply the product of combining interchangeable resources and productive labour.
>
> (De Soto 1989: 243)

The informal sector does offer a channel for entrepreneurship as it avoids the high overheads and bureaucratic problems of those of the formal sector.

Successful entrepreneurs (such as those cited by Gwynne 1985, Schmitz 1982, and Peattie 1982) all benefited from the informal sector's low overheads, cheap services and subsequent ability to produce more cheaply than the formal sector, thereby becoming more competitive within a free market situation. The vast majority of those in the informal sector, however, are the providers of cheap services. Their large numbers give rise to the unbridled competition that is so admired by proponents of neo-liberalism and also make possible the cheap reproduction of a workforce labouring to promote export-led growth. These are the informals that are not able to better their economic and social position. These are the poor that are destined to remain caught in the poverty trap.

As Portes concludes, 'The present situation is the deliberate and continuously reproduced consequence of a new worldwide structure of accumulation . . . (and) . . . an integral component of peripheral capital-ist economies' (Portes 1989: 248–9).

The rapid expansion of the informal sector is a feature that Latin America shares with other Third World areas. Calcutta's industrial development in the 1950s occurred without a corresponding expansion in regular employment. Research reveals that, as a result of this, the biggest single occupational groups were petty traders and domestic servants, while only 2.8 per cent of all earners were factory workers (Goldthorpe 1975). Much of the debate that has been discussed arose from East Africa. The earliest studies were carried out in Kenya, which led policy-makers to suggest self-employment as the answer to lack of formal-sector employment (King 1979). Gerry's study in Dakar, Senegal, reveals similar links between apparently independent petty producers and large enterprises, as exist between the garbage-pickers and the paper industry (Gerry 1979).

CONCLUSION

Motivated by a desire for a better life, rural populations have migrated to urban areas and the resulting expansion of cities has meant problems of shelter and employment. For many, the movement is one of horizontal rather than vertical mobility. Some of the discussion of the informal sec-tor and the Turner thesis suggests that, for a minority, there may be small improvements in housing or occupation, but the majority remain part of the mass of the urban poor.

Poverty arises because of the lack of remunerative and secure forms of employment. A particularly well-known example of this is the growing army of street children in Latin American cities. The international media have been quick to report the horrifying murder of some of these chil-dren, often by the forces of the state, in cities such as Rio de Janeiro and Bogotá. Driven by poverty to the outskirts of the law to survive, these children have often been made scapegoats for drug-traffickers. As the large informal sector becomes more entrenched in Third World econo-mies, the articulation between sectors becomes stronger, with the result-ing poverty for many but opportunities for few.

It is also evident that the poor are not necessarily apathetic about their situation, nor are they bound by a culture of poverty, but are quite capable of positive action to attempt to solve their problems. The construction of spontaneous housing settlements and the way this has induced several governments to support self-help schemes demonstrates the way in which squatters have taken the situation into their own hands and provided themselves with some form of shelter. The inhabitants of the shanty town have frequently achieved stability and social organisation through the establishment of personal networks and voluntary associations. The

175

informal sector has on occasion generated small businesses and creative enterprises.

In some countries, action has been taken further in the form of political protest movements. This has been especially evident in Peru and Chile, where goals have been fought for and won in this way. Although Gilbert and Ward found little sign of political protest in areas of Colombia, Venezuela and Mexico where their research was carried out (Gilbert and Ward 1985), Castells documents protest movements in Mexico City (Castells 1982). Social movements, which represent a new and increasing response of urban populations to problems, are analysed in the next chapter.

FURTHER READING

Bromley, R., Gerry, C. (eds) (1979) *Casual work and poverty in Third World cities.* John Wiley, Chichester.

De Soto, H. (1989*) The other path – The invisible revolution.* (Translated from Spanish by Abbott, J.), Tauris, London.

Gilbert, A. (1994) *The Latin American City*, LAB, London.

Gilbert, A., Gugler, J. (1992) (2nd edn) *Cities, poverty and development in the Third World.* Oxford University Press, New York.

Gilbert, A., Ward, P. (1985) *Housing, the state and the poor.* Cambridge University Press, Cambridge.

Gwynne, R. (1985) *Industrialization and urbanization in Latin America.* Croom Helm, London.

Preston, D. (1987) (2nd edn) *Latin American development. Geographical perspectives.* Longman, Harlow.

Portes, A. (1989) Latin America during the years of the crisis. *Latin American Research Review*, XXIV (3).

Ward, P. (1990) *Mexico City.* Belhaven Press, London.

Chapter

8

Social class and social movements

Chapters 6 and 7 examined the changes that have been taking place in rural and urban areas, and the effects these have had on social structure, poverty and inequality. It is now necessary to situate these in the wider context of the social formation and in particular in the context of class structure. The mapping out of social classes is valuable initially in providing a broad picture of society, but more importantly because it helps to explain the dynamics of social change. Change is invariably the result of class struggle and conflict, so that any understanding of social change requires a knowledge of how social classes act. In the analysis of change, the various theorists have focused on different agents of change, which are often social classes. Given the great variation that exists in wealth, social organisation and culture in Latin America, it is not easy to find a simple, but also heuristic schema for class analysis, and perhaps for that reason the exercise has not been attempted very often.

One of the most useful class outlines to appear is that developed by Portes, which is used here as a guide to the ramifications of class (Portes 1985). He takes as his criteria for class membership: firstly, the position of individuals in the process of production and their mode of sharing in the distribution of the product; secondly, control over the labour power of others; and thirdly, mode of remuneration. These criteria make the concept flexible and appropriate for the study of LDCs, as well as being operational. In this way, Portes distinguishes five social classes. The dominant class, having control over both the means of production and over labour power, derives its remuneration from profits and salaries and bonuses linked to profits. This is the class that enjoys both power and wealth derived from capitalist investment. The bureaucratic–technical class does not have control over the means of production, but does have control over labour power and, unlike the dominant class, finds its remuneration not in profits but in salaries and fees. This refers to the professionals and white-collar workers whose earnings are linked to their skills and qualifications rather than capitalist ventures. The formal proletariat has control over neither the means of production nor labour power, but its members earn regular wages which give them a security that the mass of the poor lack. The informal petty bourgeoisie, like the dominant class, has control over means of production and labour power, since its members own their own small enterprises. Their level of operation, however, is quite different as their irregular profits serve to keep them only slightly above subsistence level. The informal proletariat does not have the independence that control over means of production and labour power implies and must survive at subsistence level on casual wages.

How do the various classes fare in terms of participation in national wealth? This can be assessed by linking class structure to income distribution figures. Portes suggests a correlation between the dominant and bureaucratic–technical classes and the top decile in the income distribution table and also between the informal proletariat and the poorest 60 per cent of the population (Table 8.1).

Table 8.1 Income distribution in Latin America, 1960-75

	Share of total income %	
Income strata	1960	1975
Latin America		
Richest 10 per cent	46.6	47.3
20 per cent below richest 10 per cent	26.1	26.9
30 per cent below richest 10 per cent	35.4	36.0
Poorest 60 per cent	18.0	16.7
Poorest 40 per cent	8.7	7.7

Source: Derived from Portes, A. (1985) Latin American class structures, *LARR* **20**(3): 25.

In this case, between 1960 and 1975 the dominant and bureaucratic–technical classes increased their share of national income by about 1 per cent, at the expense of the informal sector, which experienced a decrease of 1.3 per cent. Portes suggests that UN agencies, using a more restricted definition of informal proletariat, would equate it with the poorest 40 per cent. This group has also lost out, its share being reduced from 8.7 per cent to 7.7 per cent.

It should be pointed out that despite increasing income concentration, there has been a general improvement in the quality of life. A Physical

Quality of Life Index, which refers to indices of infant mortality, life expectancy at age one and adult literacy shows an improvement for all the seventeen Latin American countries to which the index was applied in the period from 1950 to the mid-1970s. In 1950 the index ranged from 36 for Guatemala to 77 for Argentina, but by the mid-1970s, it had risen to a range of 43 for Bolivia and 90 for Puerto Rico (Felix 1983).

Portes also suggests figures to show the percentage of the population of each class in each country in 1970 and 1980. This demonstrates class differences between the different areas (Table 8.2). These show that the dominant and bureaucratic–technical classes together do not exceed 15 per cent of the Economically Active Population in any Latin American country and in most their proportion is much lower. The dominant class alone comprises no more than 4 per cent of the EAP in any country and no more than 2 per cent for the whole of Latin America. Despite similarities in size across the continent in these two classes, there is greater variation in the proportion of the population in the formal proletariat, but they relate to levels in economic development. In the southern-cone countries of Argentina, Chile and Uruguay, which are highly urbanised and have advanced industrial complexes, the formal proletariat makes up more than half the EAP. In Costa Rica, Panama and Peru, where economic development has taken a distinct path unlike the rest of the continent, the formal proletariat represents a quarter of the EAP. It is worth noting, however, that Brazil is not far off the quarter-mark with 20.5 per cent. Another conclusion that can be drawn from this table is that, apart from the southern-cone countries, those who work in the modern sector of the economy comprise a minority of the population. A closer look at class composition and role will put flesh on the bones of distribution statistics and also explain the extent to which the various classes have acted as agents of change.

BENEFICIARIES OF THE SOCIAL ORDER

Traditionally the dominant class had its basis in landownership, but over the years these élites have often shifted their capital to the urban sector because of the possibility of better returns on it, and, on occasion, because of agrarian reform. Today the number of major companies that are privately owned by Latin Americans is small, because important sectors of industry, agriculture, mining and commerce are controlled by foreign-owned or state enterprises. Top earners in Latin America may be owners or managers, they may work in foreign or local firms and they may work in state or private enterprises. The group that Portes subsumes under the heading of a dominant class are internally divided, but have enough in common for Portes to call them a class.

Potential divisions within the class are numerous. The dependency

Table 8.2 The Latin American class structure[1]

Country	Dominant 1970 (%)	Dominant 1980 (%)	Bureaucratic technical 1970 (%)	Bureaucratic technical 1980 (%)	Combined dominant and bureaucratic-technical 1970 (%)	Formal proletariat 1972 (%)	Informal petty bourgeoisie 1970 (%)	Informal proletariat 1970 (%)	Informal proletariat 1980[2] (%)
Argentina	1.5		7.5		9.5	59.0	9.7	22.3	23.0
Bolivia	1.3	0.6		5.7	5.7	3.3	4.8	86.2	56.4
Brazil	1.7	1.2	4.8	6.4	10.2	20.5	7.2	65.8	27.2
Chile	1.9	2.4	7.1	6.6	7.7	60.5	4.5	26.0	27.1
Colombia	0.7	0.7	4.5	4.3	6.6	12.9	15.7	66.2	34.3
Costa Rica	1.7		8.0		9.0	28.5	13.5	48.3	19.0
Dominican Republic	0.3	0.4	2.7	3.1	3.7	6.4	17.3	73.3	
Ecuador	0.8	1.0	5.0	5.1	4.7	10.0	4.1	80.1	52.7
El Salvador	0.2	0.5	3.0	4.2	3.8	5.2	23.1	68.5	39.8
Guatemala	1.6	1.1	3.1	3.7	4.5	22.3	3.3	69.7	40.0
Haiti	0.5		0.5			0.0	[3]	[3]	
Honduras	0.6		2.5		4.5	1.1	13.4	82.4	
Mexico	2.6		6.2		7.7	15.9	11.3	64.0	35.7 ·
Nicaragua	0.9		5.2		5.3	8.7	15.8	69.4	
Panama	2.1	4.4	6.8	10.0	8.7	25.4	5.2	60.5	31.6
Paraguay	0.6		4.2			5.9	[4]	[4]	

						[5]		
Peru	0.4	7.6		7.0	27.6		69.5	40.4
Uruguay	1.1	5.6	7.3	8.4	88.5	1.0	3.8	
Venezuela	3.6	8.6	9.5	10.0	12.2	14.0	61.6	20.8
Latin America[6]	1.7	5.4	6.0	8.4	22.4	10.2	60.3	30.2
(N = 93 850 000)								

Source: Adapted from Portes, A. (1985) Latin American class structures. *LARR*, 20 (3): 22–3.

[1] All figures are percentages of the domestic EAP. Figures are for 1970 unless otherwise indicated.

[2] Percentage of the non-agricultural EAP represented by unremunerated family workers and the self-employed.

[3] Insufficient data to differentiate between the informal petty bourgeoisie and the informal proletariat. The total figure for the two classes is 99.0.

[4] Insufficient data to differentiate between the informal petty bourgeoisie and the informal proletariat. The total figure for the two classes is 95.2.

[5] The estimation procedure yields a negative figure in this case.

[6] Averages weighted by the proportion of the regional EAP in each country.

181

theorists logically identified the main distinction as being between the national bourgeoisie, whose enterprises were based on local capital, and the *comprador* bourgeoisie, whose interests were tied to foreign capital. They initially thought that competition from powerful foreign interests would encourage the national bourgeoisie to take an anti-imperialist stand. It was soon realised, however, that the national bourgeoisie was too weak to pursue an independent line and being themselves involved in the complexities of dependent development, their orientations remained the same as the rest of the dominant class. In Argentina, for instance, the national bourgeoisie has various indirect links with foreign capital. Many local industrialists rely on credit from international loan agencies, others import essential supplies and the majority import essential equipment. Petras and Cook found that only 4.9 per cent of their sample of the top national bourgeoisie were not dependent on technology that originated from abroad. These industrialists welcomed foreign capital, because they felt that it stimulated economic activity in general, from which they could all benefit (Petras and Cook 1973). In Peru, up-and-coming executives use a period of employment with a multinational as a form of training in the methods and values of big business and then may shift to employment with state or private Peruvian firms or go into business on their own (Becker 1983).

Other commentators have drawn attention to the basic distinction between owners and managers, which is important for a strict Marxist analysis. This distinction, however, carries less weight where managers of multinationals have control over much greater economic resources and far more people's livelihoods than the owners of small local firms.

One would also expect a difference in attitudes between general managers of state corporations and those of transnationals. The expansion of the state in the economy has led to an increase in the number of state employees, some of whom direct vast industrial complexes. The top personnel in these enterprises have been referred to as a state bourgeoisie. We can, therefore, identify three fractions: domestic capitalists; managers of multinationals and top administrators of public enterprises.

What the different groups all have in common is their control over production processes in the economy and over the labour of a number of subordinates. Their remuneration comes from this position and takes the form of profits or high salaries and bonuses tied to profitability, so that their financial reward is linked very closely to the success of the firm. Since all the members of this class wish to preserve the status quo, they unite politically to support conservative regimes and oppose radical forces that propose significant social changes.

Evans argues that the three fractions have a common interest in capital accumulation and in the subordination of the mass population, and therefore cooperate in what he calls a 'triple alliance' between élite local capital, international capital and state capital. Local capital is not assumed subordinate because the local bourgeoisie has certain economic and political advantages, which can be used when dealing with multi-nationals. There is some competitiveness within the alliance, but this is reduced by bargaining and a working cooperation is maintained (Evans 1979).

The main feature of the bureaucratic–technical class is that although it lacks effective control over the means of production its members do have control over the labour of others in their subordinate position in bureaucratic structures. Unlike the dominant class, remuneration is tied to specific salaries and fees rather than the profitability of a firm. In Latin America, this class is composed mainly of middle-level management and technical personnel in domestic, state and foreign enterprises, and functionaries of the state bureaucracies, including the armed forces and independent professionals. This class maintains the smooth running of the social order and economic system, but differs from the dominant bourgeois class in that its wealth is not derived from capital. With the dominant class, it shares in the control of the existing social order and benefits from that order (Portes 1985).

The dominant and bureaucratic–technical classes benefit from the system, but how do they act in this privileged position? Do they initiate change and if so, what kind of change? Although the terminology and implications differ, both traditional Marxists and non-Marxists have attached a progressive role to the new entrepreneurial classes that emerge with industrialisation. Marx saw them as promoting the bourgeois state, which was an advance on feudalism and which, in turn, would give rise to the proletarian state. Modernisation theorists, leaning heavily on the experience of Western Europe argued that the middle class erodes the traditional order and promotes economic, political and social progress. In the Latin American context, Johnson took the lead in proposing the middle class as the spearhead of change, by detailing the areas in which he considered they had influenced government in a progressive way in five major republics (Johnson 1958). Empirical research in the 1960s demonstrated that even in the cases where the middle class had been initially progressive, success had given them a conservative outlook and led them to join forces with the traditional élite rather than with the working class (Ratinoff 1967; Sunkel 1965).

A closer look at the bourgeoisie, in terms of their social background, relationships with other classes and political opinions, lends support to the argument that, in practice, they are not the class that is going to bring about radical change. Entrepreneurship has not been a significant mechanism for social mobility in Latin America. Several studies have shown that successful Chilean industrialists since the 1930s, before which immigrants dominated the industrial scene, have come from land-owning or well-to-do families (Johnson 1972; Cubitt 1972). Research on Colombian entrepreneurs reveals a similar pattern, with only 9.5 per cent coming from working-class backgrounds. What is particularly interesting about the latter findings is that four-fifths of the socially mobile were born outside Colombia. Lipman points out that it is easier to gain the necessary qualifications and training for success in Europe. This would suggest that social mobility within Colombian society is more limited than the figures imply (Lipman 1969). The country that has demonstrated the greatest upward mobility through entrepreneurial channels is Argentina, which was one of the first to industrialise and which has had probably the greatest influx of immigrants. Imaz found that about a quarter of the Argentine industrial élite, at a time when industrialisation was getting well under way, were self-made men from neither middle- nor upper-class backgrounds (Imaz 1964).

Explanations for this lack of mobility lie in the international context. Initially, the immigrants provided too much competition for the aspiring local working class, as, coming from countries where the process of industrialisation was under way, they were able to use their knowledge and experience to seize the initiative in entrepreneurial activity.

More recently, multinationals and foreign capital, with all their implications, have made vertical upward mobility difficult. The promising entrepreneur finds it very hard to compete with multinationals, which can always outproduce him and undercut his prices because of the very scale of their operations and capital. To compete successfully with multinationals, the entrepreneur must acquire knowledge, qualifications and specialised training similar to the managers and technical staff that the foreign company can afford to employ. Higher education is expensive in Latin America, and especially the sort of education, such as business qualifications from the USA, that is desired for running large firms. Only the offspring of the wealthy can afford to benefit. These factors again demonstrate why the wealth produced by productivity, on the basis of foreign capital and in a situation of unequal opportunity of access to qualifications, does not produce the desired trickle-down effect.

Given their social background, it is not surprising that the bourgeoisie, in most Latin American countries, has close links with landowners. Research in both Argentina and Chile has revealed the close links between industrialists and agrarian interests (Chilcote and Edelstein 1974; Petras 1969).

In the case of Chile, it is argued that the successful new urban groups have been coopted by the land-owning oligarchy. According to Sunkel, the major changes have been industrialisation, urbanisation and an expansion of the economic and social activities of government which have, together, brought an improvement in the standard of living of urban middle-sector groups. Chilean social structure has always had a small élite, with a great concentration of power and, instead of presenting a challenge to them, the successful industrial and commercial groups have become incorporated into their ranks. This social fusion has taken place through intermarriage and access to the social circles and institutions which denote prestige in society, such as organisations, clubs and landed estates (Sunkel 1965).

This point is supported by Johnson, who goes on to demonstrate that, as a result, Chilean industrialists have not developed their own independent ideology and value system. Questions about their politics elicited answers which were generally conservative and not in favour of progressive ideas. Johnson found little support among them for the reformist policies of the government of the day. Less than a quarter were in favour of the government having the power to expropriate land, which offers little support to the opposition to the landed élite thesis (Johnson 1972).

Considerable evidence shows that, to date, those who have been successful through industrialisation and urbanisation have not been an independent, dynamic and progressive force in Latin America. Why should this be so? Chilcote and Edelstein argue that it is because of the type of industrialisation that has taken place. Industrialisation was prompted by the economic weakness of Europe and the USA during the Great Depression and the Second World War, leading to import substitution rather than a conscious development policy. The implication is that industrialisation was not the planned result of a dynamic, forward-thinking group but a reaction to a crisis (Chilcote and Edelstein 1974). The point is that so much economic growth in Latin America, from the enclave economy to contemporary multinationals, has been brought about by foreign concerns, which has denied the local entrepreneurial groups their historic role in leading, organising and financing this process. The only time they had the opportunity was, almost by default, during import substitution, but Latin Americans were soon overshadowed by the transnationals. Since so much capital has external sources, the bourgeoisie have not come through the same phase of saving and investment as did their European counterparts, but have moved straight to a consumption stage. One argument states that the Latin American bourgeoisie has never had sufficient capital to promote the economic development it wanted and has, therefore, had to import capital. The counter-argument is that it is precisely the importation of capital that has prevented the local bourgeoisie accumulating its own on a grand scale.

As a possible agent of change the bourgeoisie was discarded by dependency theorists, but in the wave of post-dependency criticism, there have been attempts to resuscitate this role for them. Becker says that it is expecting too much of Third World bourgeoisies to think they can bring about societal transformations on a grand scale or act in the 'heroic' way of entrepreneurs a century ago (Becker 1983). He characterises the mining bourgeoisie in Peru as a corporate national bourgeoisie, which he considers to be progressive as it is nationalist and developmentalist. This new bourgeoisie is innovative technologically; internationalist, since its members partake of universalistic norms of technocracy and the managerial ideology; and it promotes development, through a desire to compete with transnational companies. The group is strong because it has access to power through its institutional involvement with the state, which is why Becker calls it a corporate bourgeoisie. Becker asserts that greater social mobility is possible in a society where the new corporate bourgeoisie is the dominant class than where family firms predominate in the economic sector. The conservative aspect of the new bourgeoisie is that it establishes a more durable form of capitalism, but as Becker does not see socialism as the only way of ushering in progress, this does not present a problem for him in his description of the new corporate bourgeoisie as promoters of development.

THE WORKING CLASS

The formal proletariat is defined by its lack of control over both the means of production and the labour of others (Portes 1985). Its remuneration in the form of wages may not be very high, but does offer a measure of security, in that wages are contractually established and regulated under existing labour laws, which means that in theory they can not be arbitrarily withdrawn or altered. The formal proletariat is basically urban in composition and along with the dominant class and the bureaucratic–technical class, makes up what is referred to as the modern sector of Latin American economies. The Latin American working class has its similarities to and differences from the working class in other Third World areas. Where the nature of development has been capital intensive in LDCs, the proletariat will be small. The previous chapter showed how so much of the rural–urban influx has been absorbed by the informal sector rather than regular working-class employment. Even if development is not capital-intensive, but relies on labour power, what is emerging in Latin America is a pattern of part-time employment, whether it be seasonal work in rural areas or informal-sector activities in urban areas. To survive in this situation it is necessary to have more than one economic activity. Though small, the working class in Latin America, because of higher levels of urbanisation and industrialisation, is relatively larger and better established than in most other Third World regions. The pattern of increasing underemployment, both rural and urban, may be one that will be followed by the other developing countries.

Figures for the proportion of the total population in the labour-force are quite revealing. In the mid-1970s, while almost half the population of the UK and the USA (47 per cent and 45 per cent respectively) made up the labour-force, in Mexico (which had a very young population) only 28 per cent were working. Chile had a similar 28 per cent and Peru 29 per cent, while Venezuela and Brazil fared a little better with 31 per cent and 32 per cent of the total population in the labour-force (World in Figures

1978). What this means is that only a relatively small proportion of the population is earning and so has the burden of supporting a large number of old and very young. Portes, using statistics for 1972, has assessed the proportion of the Latin American economically active population that constitutes the formal proletariat as being 22.4 per cent (see Table 8.2).

Although the working class is of recent formation it operates in a variety of different working situations. The Latin American economy is characterised by a large number of very small firms and yet some workers are employed by large transnationals using the most modern technology and employing thousands of workers. Where mineral-extracting enclaves still exist, the workforce lives in its own community, a long way from major urban centres. The various levels of technology employed also mean that workers have very different skills and qualifications.

It has been argued that these divisions within the working class are irreconcilable and prevent it acting as a strong, united, radical force. The proposition put forward by this view is that technology shapes the problems, demands and forms of organisation of workers, and it is precisely the levels of technology that vary so much between the modern, dynamic, industrial sector and traditional areas. The better-off within the working class have often been referred to as a labour aristocracy, though the term has been used differently by various social scientists. Some use it generally to refer to those workers whose income is higher than most others. Arrighi and Saul more specifically consider the labour aristocracy to comprise the skilled technicians in large factories and sub-élite of clerks and teachers (Lloyd 1982). Others identify the labour aristocracy as workers who have high earnings because they are employed by multinationals or foreign firms which, as Roxborough points out, is closest to Lenin's original meaning (Roxborough 1979). The implication of the term is not only that it is a privileged group but that, as a result, its members will not desire change and so be a conservative force in society. The two issues at stake with the labour aristocracy thesis are: can this relatively privileged sector of society be a militant force for change or will it want to maintain the status quo? Can this group remain within a united working class or will it pursue selfish interests, separating itself further from the remainder of the working class?

Humphrey's research on workers in the Brazilian motor industry in the 1970s shows that, in this particular case, a group of relatively highly-paid workers in a modern sector of the economy did not behave as a labour aristocracy, but took on a vanguard role in providing political leadership for the working class as a whole (Humphrey 1982). He also points out similar situations in Argentina in the 1960s, when car workers were involved in revolutionary unionism and urban insurrection, and in Mexico, where they have fostered some of the stronger, independent unions.

In the Brazilian situation, it was the car workers who took the leading role in the industrial disputes of the 1970s and early 1980s. They took full advantage of the general liberalisation that was the government's policy at the time to fight against the constraints imposed by the labour system. Despite their relatively high wages, these constraints had imposed burdens on working conditions. Through their militancy they won union autonomy and union reform. Humphrey says 'in the seventies, at least, factors which distinguished auto workers from other sections of the working class enabled them to adopt a vanguard role' (Humphrey 1982: 231). Because they worked for large, modern firms, the state was closely involved with their employers, which meant their industrial grievances led them into conflict with the state. To challenge their employers, they had to challenge the labour system itself. Their activities were supported by, and benefited, the rest of the working class. Support can be seen in the upsurge of working-class activity that accompanied the strikes of 1978 and 1979 and the way that the union's leader, Lula, became such a popular figure. The union reform that was won by these workers benefited the working class in general.

Humphrey calls the differentiation of the working-class thesis technologically determinist because he says it attaches too much importance to technology. The development of industry is part of a process of the accumulation of capital that involves not only technological changes, but also other factors. The working class is affected by the changing relationship between labour and capital, the changing role of the state and new forms of labour legislation. The aim of management is to make profits and, therefore, its strategy with the labour-force is directed to this end, not dominated by technology. The factory, too, is not solely a technical institution, it is a social organisation, influenced by political, social and economic aspects of society at large.

Although Humphrey emphasises that one can not rule out the possibility of car workers being a conservative force in the future, he does demonstrate that at this particular historical–political conjuncture their militant action provided a catalyst for change for the working class. The political stance of the working class will rely heavily on the nature, size and development of the labour movement. It is the union movement which has, in the industrialised countries, been the basis of action. A look at the development of the labour movement in Latin America will lead us into the debate over the radical/conservative nature of the working class.

Spalding has divided the period of growth of the labour movement into different stages (Spalding 1977). The formative period from 1850 to 1914 witnessed the birth of a clearly identifiable movement, with labour organising for the first time on a massive scale. Prior to that, guilds had been the only groupings of workers. In Argentina, the formative period began with the formation of mutual aid associations for shoemakers and printers, and ended with the foundation of the General Workers' Union and three general strikes. The period from 1914 to 1933 Spalding identifies as the expansive and explosive period. This was a time of peak union activity for many countries, for both qualitative and quantitative advances were made. New ideologies from Europe provided workers with different organisational concepts. In the southern-cone countries, the union movement, having spread to white-collar workers and peasants, erupted in large-scale worker protest.

All this union activity did not occur without any reaction from the élite. For them, the expansion of the labour movement was perceived as a threat of communism, so various measures were resorted to in order to control labour, ranging from crude repression to legal limitations on union activity. At the same time, many governments offered social reforms to ease the plight of workers and give them fewer grounds for opposition. This labour legislation included laws governing female and child labour, improvements in working conditions and social security provisions. But, even where there were legal improvements, enforcement was often limited and, given the nature of the labour market in Latin America, many workers remained outside the legal framework, which applied only to industrial workers.

Spalding calls the period from 1930 to the present day the cooptive period, because of the attempts made by the élite to integrate labour into the established political and social framework. By this time, politicians had realised that organised labour represented a bloc of votes which could be very valuable to candidates. Control from above was won by the use of various mechanisms. Ideologies, such as nationalism and developmentalism, were used to counter the influence of communism. Nationalism encouraged workers to identify with their nation rather than their class, while developmentalism asked the worker to sacrifice short-term gains, such as better working conditions and remuneration, for long-term interests, on the basis that industrialisation benefits all in the long run. Labour codes as a mechanism of control were introduced in most countries. These included (1) the power to intervene in labour federations and unions; (2) control over candidates in union elections; (3) provisions governing the legality of strikes and unions; (4) legal distinctions between white- and blue-collar workers which have the effect of dividing workers; and (5) the prohibition of certain groups (often government employees) to form unions. Spalding also suggests that the introduction of highly technological industry, leading to increased levels of unemployment, has weakened the labour movement and made political alliances necessary.

Spalding concludes that, with few exceptions (such as the tin miners in Bolivia), labour has been coopted by these mechanisms and not proved itself to be a revolutionary force. It is true that the very process of mobilising labour by élites for their own support has brought workers together for political action and this could form the basis for the development of an independent labour movement, but this has so far been prevented by the mechanisms of control (Spalding 1977). Spalding's stages could be criticised for being somewhat vacuous. The term formative is fairly obvious for the beginning of a development process so, also, the term expansion for the next stage. Nor do the stages neatly fit all the Latin American countries. Nevertheless, they do provide a rough scheme for organising the multitude of empirical facts concerning the growth of unions. What is more polemical, is the conclusion that the labour movement, despite periods of radicalism and fervent political activity, has become a conservative force in Latin America.

THE ROLE OF THE PROLETARIAT DEBATE

For Marx, class consciousness, which develops with the increasing struggles that the workers have with management, is a key factor contributing to the revolutionary role of the proletariat. The supporters of Spalding's view would argue that in Latin America the working class has not developed into a fully-fledged class in its own right. Much of this can be explained by the cooptation which has taken place through the mechanism of clientelism. As with the peasantry, strong ties of loyalty and obligation tend to prevent the development of permanent horizontal links. These may shape the activities of unions, so that their political manoeuvring is based on clientelist relationships rather than class relationships. Payne argues that in Peru this method has been very successful, for by using it, the organised workers realised a greater improvement in wages than the non-organised white-collar workers, and achieved a rise in their real income, even in the face of inflation (Payne 1965). These tactics have been used by unions in Mexico, where cooptation through the state bureaucracy has taken place as the labour movement is integrated into the organisation of the governing party. Mexico's electoral system has given power continuously to the party that emerged triumphant from the revolution, the Revolutionary Party or PRI. One of the three major sectors of PRI represents labour and within this, the CTM (the main confederation uniting Mexican unions) predominates. The CTM has a complex organisation of state-wide regional and local federations, based on geographical cohesion and industry-wide unions on a national and regional level. In any industrial plant there is a local union headed by a general delegate. The close integration of labour with the ruling party means that labour has no independent base. It assumes that the interests of workers are national, rather than class interests, which may not always be best for the workers, as a class. Mexico is often depicted as being characterised by '*charismo*', the phenomenon of trade unions being controlled by the state, in order to keep wages down in the service of capital accumulation and accelerated economic growth. It is frequently argued that the political stability and rates of economic growth of the post-war

190

period rest on the control of subordinate classes by the Mexican state. Since the union organisations are part of PRI, they have a dual function: firstly, as a pressure group lobbying for a greater share of social benefits for labour; secondly, as an apparatus of political control of the working class. These functions are coordinated through clientelism, which exists between local leaders and their fellow members, and also between local leaders and national and regional leaders, with the local leaders this time being the clients. In San Cosmé, as elsewhere in Mexico, union leaders have an important say in job placements in factories and can exert influence on behalf of their members at regional and national levels. Control of jobs and access to political leaders can be used to win political support for themselves. In the role of clients, these local leaders bring in their own supporters behind their patron, in return receiving political advancement for themselves and small benefits for the union (Rothstein 1979). Lomnitz argues that this situation is exacerbated and class identity weakened by the lack of informal ties and solidarity between workers. She says that as Mexican workers socialise mostly with the extended family, there is little personal contact with co-workers (Lomnitz 1982).

The linking of the national President of Mexico with the workers through clientelist relationships can be seen in the pattern of strikes over the years. Gonzalez–Casanova, using statistics for numbers of strikes per year, suggests that these follow the presidential-type policy of the government of the day. He found that the greatest number of large-scale strikes broke out when the presidents were known for their populist and pro-worker policies. This implies that workers and union leaders felt themselves protected and even encouraged by the presidential power. Presidents who followed less radical policies or more open alliances with the property-owning national and foreign sectors they felt to be less sympathetic to their cause, and therefore less likely to respond to their wishes. It was clearly patronage that they were seeking, rather than a confrontation with another class. The unions had sought and relied on government policy, rather than developing mass working–class organisations characterised by class-consciousness (Gonzalez–Casanova 1968).

Roxborough criticises the standard account of the Mexican labour movement for its one-sided emphasis on the aspect of control over the rank and file (Roxborough 1984). He argues that the extent of this control has been exaggerated, and as evidence documents a number of cases of successful, although sometimes short-lived, insurgent labour movements that have taken place. He found in his research on Mexican unions that there was no direct link between union affiliation and militancy: in other words, the unions that were controlled from above were no less likely to be militant. This is not to say that the Mexican unions are likely to become revolutionary tools to be used against a capitalist state, but nor are they conservative in supporting the status quo. Roxborough believes it unrealistic to expect unions to play such a revolutionary role in Latin America today and predicted that with the economic crisis of the 1980s in Mexico, the unions would become more militant (Roxborough 1984). This in fact has not been the case. They have been undermined both by new government strategies and the growth of social movements.

From the point of view of Spalding and Gonzalez-Casanova, the working class is seen to have been coopted by the political élite, who have offered them favours such as improved wages or working conditions, receiving in return political support in the form of votes of union members. For the workers, this strategy has not been entirely unprofitable, and indeed in Peru (Payne 1965) and Mexico workers have received some definite material rewards through it. There is no doubt, too, that the time during which the working class did best in Argentina was the Peronist period, when clientelism linked the labour-force with the powerful. In 1954, the last full year of the Peronist era, the wage-earners' share of the national income was 30.6 per cent, an all-time high (LAWR 1983). Peron was a political patron at the highest level, who offered his workers definite improvements in return for their support on which he depended.

Although these mechanisms may provide short-term micro-level benefits, they do not encourage the development of an independent organisation with its own ideology, which would then be in a position to spearhead change to bring about real and long-term improvements. Favours are won through reciprocity, not through militant industrial action.

Contrasting views, based on different areas, portray the working class as having greater potential for radical action. Lloyd points out that it is often those who are called a labour aristocracy, who are the most militant. 'It is they who are most exploited and they are conscious of this. Bargaining, however, tends to be confined within the company. Selfishness is also seen in their apparent lack of concern for the poor but what, in effect, could they do?' (Lloyd 1982: 119). Among the African poor this militancy is often seen as leadership, rather than being viewed antagonistically.

Argentine unions, which were shaped from above, are some of the strongest in Latin America. Peron needed the union movements for support and, indeed, it was the labour movement who brought him back from exile and won him the presidency. Workers' organisations were then created and expanded, with the assistance of patronage, but in order to defend workers' interests. When Peron was overthrown, these organisations were strong enough to resist opposition. To this day, despite periods of extreme repression, they have survived, often having to spend long periods underground.

The Bolivian tin miners and the Chilean unions have demonstrated that it is possible for the labour movement to develop a revolutionary role in Latin America. The Bolivian miners performed the classic vanguard role when they participated in the Bolivian revolution and became significant partners in the revolutionary party that took control. Today, however, the power of one of the most militant movements in the region has been undermined by the decline of the tin market (Martin 1986).

Along with Argentina, Chile was one of the first countries in Latin America to experience union organisation, and by 1970 the labour movement had emerged as a strong and ideologically independent organisation. The new left-wing government, under the leadership of Allende, restructured the economy into three sectors: (1) the social-property sector (made up of state enterprises and those to be expropriated); (2) the private-property sector; and (3) the mixed-property sector (made up of enterprises where the capital was owned jointly by the state and private individuals). The workers, wanting a say in management, exerted pressure on the union leadership and the government (who came to an agreement in December 1970) and so ensured their participation in the running of enterprises in the social and mixed sectors of the economy. Concrete modes of putting this into operation were discussed by a commission and put into effect by June 1971. The expropriated firms were run by an administrative council made up of eleven members, five elected by the workers, five appointed by government, and a director appointed by the President. Of the workers' representatives, three were elected from the shop floor, one from the administrative workers and one from the technicians, engineers and graduate staff.

Worker participation was a notable achievement, but the Chilean proletariat went further in their radical actions, by setting up the *cordones*. These were the result of spontaneous revolutionary activity from the base, which even embarrassed the union leadership, and demonstrated a high level of class consciousness among the mass of workers. The *cordones* were units of industrial organisation made up of workers from various enterprises, which were brought into being by the Bosses' strike in October 1972. A left-wing party outside government had for some time been calling for the workers to seize factories; thus, when they found themselves shut out by the closing of firms by management in the Bosses' strike, the workers were prompted to take action. They seized firms and, having taken them over, set up a *cordón* in the area. They made an inventory of the industries in the zone and then contacted union members in them. It was frequently these groups that were responsible for worker takeovers, for if any of the firms in the area were still in the private sector the local *cordón* would encourage the workers to take control. One such *cordón* was the Cordón O'Higgins, which in 1973 grouped fifteen enterprises and, with a base of 11 000 workers, mobilised 5000 of them.

This activity worried the union leadership and the government because, although they were committed to a strong social-property sector, the government was adamant that this should be achieved by legal and orderly means. The government was also committed to a private sector as in its manifesto it had pledged support for the small business-man, of whom there were many in Chile. Legality was important to the President because, although he had won the presidential election, he did not have a majority in Congress. He, therefore, could not afford to support illegal methods which would give the opposition in parliament and the middle class in general genuine grounds for opposing him.

The Chilean case shows that the working class was prepared to take the situation into its own hands and adopt radical action over the heads of its own leaders. Highly organised and ideologised, and prompted by a sympathetic political situation and finally by the political actions of management, it became a powerful militant force. Chile's previous period of democracy, and the way in which the labour movement had developed its independent, political base and strong organisational structure during that time, made possible the development of militant collective action on the part of the proletariat.

'The political stance adopted by a class at any given time will be in part a function of the structure of the political system as a whole and the concrete possibilities which exist in a specific situation for the application of various kinds of class alliances' (Roxborough 1979: 82). This is true of the working class in Latin America, whose political stance has varied across the region. Most of the evidence from Mexico lends support to the Spalding view, while the data on the Allende period in Chile show very distinct revolutionary potential. Humphrey stresses that his conclusions relate only to the Brazilian car workers under study at that particular time. The next chapter looks at the way in which the working class has been involved in class alliances in order to capture the state and so exert its influence.

POPULAR SOCIAL MOVEMENTS

Perhaps it is unrealistic to expect any class to take on a revolutionary role in contemporary Latin America and more fruitful to recognise the advances, however small, made in one particular place and time. The previous sections have shown that these have occurred, but that social classes in general have not been effective agents of structural change. In this context, new attempts at change in the form of the activities of popular social movements have been emerging since the 1970s.

In the early days of shanty-town development, there were suggestions that the squatters would become a strong political force with the potential for threatening the status quo. Land invasions are a highly militant form of action which require courage on the part of those involved. But in most cases, the land invasion represented the peak of political activity. In general, further energies went into consolidation, as the struggle for day-to-day survival offered little time or opportunity for political mobilisation. Exceptions to this were the case of the Chilean urban poor before and during the Allende period and Peruvian collective action during the 1960s and 1970s. Although this sector has rarely produced a highly organised and successful political movement, it is becoming more of a source of protest that is gaining significance in contemporary Latin America. Popular protest movements have become more common, and although they are usually multi-class in membership they often originate in poor urban neighbourhoods.

These social movements express new forms of social struggle which have arisen out of the relative failure of other sectors of society to make significant changes and out of changing socioeconomic conditions. They are not confined to Latin America, nor even to the LDCs, as they represent a common response to some of the problems of contemporary capitalism. Three major processes are identified as being responsible for new forms of subordination that inspire resistance in the form of new movements: firstly, a commodification of society as individuals are increasingly brought into a market system; secondly, bureaucratisation resulting from increasing state intervention into daily life and finally cultural massification as the influence of mass media becomes more pervasive (Slater 1985).

The term social movement refers to a wide range of groups with certain characteristics in common. Firstly, they are multi-class, which does not mean that class struggles do not exist. These are new antagonisms which emerge as social conflict is diffused to more social relations. Secondly, instead of being located in a work situation, these groups are usually based in a community, often a neighbourhood or some form of local grouping. They bring together people with a shared experience of suffering the same problems. Thirdly, social movements are organised around specific demands such as the need to defend the legality of landholdings; the desire to get access to water and electricity; or the need to protect human rights. Fourthly, the movements are based on a principle of democracy, so that they have very little in the way of hierarchy or authoritative positions. Finally, Portes suggests that during the struggle the groups rarely come into direct confrontation with the dominant class, but usually with members of the bureaucratic–technical class who staff the agencies of the state (Portes 1985).

There are a number of different groupings, which may be included under the rubric of social movement: neighbourhood councils, which include the gamut of *barrio* and community organisations aimed at solving problems of urban facilities and also those that are concerned with living standards, particularly in the form of rising prices of food, transport and other necessities; women's movements, which have become particularly prominent in Grenada and Nicaragua, but which also exist in more localised forms in other countries; human rights groups, which have become so well known on a national scale in Argentina, but which are also active at neighbourhood level (as for example in São Paulo, where one was formed in 1978 in the barrio of São Miguel after one of the inhabitants was victimised by police brutality: Singer 1982); CEBs, as described in Chapter 4, are spreading now throughout the continent; regional movements have emerged in countries where the interests of one region have been subordinated to those of other areas as in Peru; and political protest groups, which in the extreme become guerrilla groups, such as Sendero Luminoso (Shining Path) in Peru.

The upsurge of social movements in the late 1970s and 1980s stemmed from both economic and political factors. The housing shortages and lack of urban facilities created by rapid urbanisation along with the economic crises of the period exacerbated the problems of the poor and heightened their awareness of the inability of existing structures to solve them. In Mexico the 1982 crisis, together with the 1985 earthquake and the government's austerity measures to service the national debt, pushed the poor to an even more marginal level of survival and encouraged the growth of CONAMUP (National Council of Urban Popular Movements). Founded in 1979 as the umbrella organisation uniting the network of local and regional popular movements, CONAMUP grew in influence as the economic conditions worsened. The Chilean movements that began in 1983 and expanded in the very poor areas were encouraged by government policies of deindustrialisation that led to chronic unemployment. These organisations focused on demands for subsistence, housing and utilities.

The failure of political parties and unions to solve any of the mounting problems led to a disillusion with existing political structures. The Chilean dictatorship eliminated most mechanisms of mediation between government and population by repressing political parties and unions, thus leaving the poor little alternative but to form their own popular movements.

Mainwaring has argued that social movements flourish under authoritarian regimes precisely because there are no alternative forms of expressing political opposition. Since formal opposition parties or organisations can not operate in any meaningful way, in order to protest and attempt to improve the situation, the population can only turn to informal groupings. Mainwaring has shown that in Brazil social movements were most active and influential during the years of the military, although reaching a peak in the period from 1974–79 when the government initiated a cautious liberalisation process that allowed greater space for popular organisation. Since democratisation the process has become more complex with some new movements emerging while others decline or disappear (Mainwaring 1987).

A criticism of this argument could be made, given the number of movements that have gained in strength under democratic governments. So many of the members of these movements live outside the formal economy that even under democratic regimes they play little part in formal political groups.

Most of these movements began with a grass-roots response to problems of poverty, as did the Cost of Living Movement in São Paulo (Singer 1982). It originated in the Mothers' Clubs in a parish in the southern part of the city. The members wrote a letter to the authorities protesting at the increases in the cost of living, which had exacerbated problems of nutrition, health and transport. After the letter was published in the press during some local elections, the inhabitants decided to continue with the campaign by carrying out research on the effects of these rises on life-styles. Seventy Mothers' Clubs distributed 2000 questionnaires, which showed that the average worker in that part of São Paulo earned one and a half minimum salaries, which was not sufficient to cover costs of food, rent, mortgage of land payments, clothing, transport, medicine, school equipment, electricity and water for their families. This was followed by a demonstration to which the Governor of the state was invited but did not attend. Various suggestions were made but little was done, leading to disillusionment which almost ended the movement. The following year, however, it was revived with the aim of extending it to other neighbourhoods of the city. Gradually the movement spread until it was able to assemble 5000 people with a petition for which almost 1 300 000 signatures had been collected, but the petition received little attention from the authorities. Although the petitioners realised that it was government policies which were responsible for the impoverishment of the masses, the movement did not oppose government, it appealed to it for improvements. Although the MCV failed in its immediate objectives it can claim some achievements. It had managed to reach the mass of people and make them aware of possible means of tackling their problems. Consciousness-raising of this type has spread, as the MCV has emerged in other cities and in other states.

A distinction can be made between those movements concerned with strategy and those centred on the notion of identity. The former, involving resource mobilisation, refer to most of the movements mentioned so far in this chapter, and require theories that feature concerns of strategy, participation, rationality, expectations and interests. The latter, which include some of the women's movements noted in Chapter 5 are analysed in terms of identity-centred theories, which 'emphasise the process by which social actors constitute collective identities as a means to create democratic spaces for more autonomous action' (Escobar and Alvarez 1992: 5).

The new social movements represent a very positive force: they unite people in different occupations to confront a problem and, in so doing, they raise consciousness. Because they operate at grass-roots level, they involve people in the issues that really matter to them, which may be the reason why they have often been so dynamic. The problem is that the factors which contribute to their dynamism are also disadvantages. Because they initially focused on local-level, small-scale issues, their achievements must by definition be limited to that level, and success (i.e. the accomplishment of the task) may even mean the disbandment of the movement. Also the lack of organisation, although contributing to the involvement of members, makes it difficult to expand and develop these movements to any sort of state or national level. In some areas this difficulty has been overcome by movements uniting to form umbrella organisations or alliances being formed between them and other pressure groups. Where Mexican groups have been successful in improving urban facilities or acquiring plots of land on which to build, that achievement can clearly be attributed to CONAMUP. It is the cooperation between the member groups and their joint numerical strength that give CONAMUP its political clout.

Some of these movements, such as the Guadalajara-based OICO (Independent Organisation of Settlers in the East) have become highly organised in the way in which they operate (Craske 1993). This group was originally founded by a number of dissatisfied residents and some seminarists sympathetic to Liberation Theology. It is now part of a complex structure that organises popular mobilisation in the city and has fifteen constituent groups. Representatives of the latter meet weekly to discuss strategy and news. As well as their political project, OICO is involved in a self-help construction group, a group to organise a kindergarten, Bible study and a parents' group.

It is significant that the Mexican Solidaridad programme of the early 1990s has been based on the notion of community action and power. This programme, inaugurated by President Salinas de Gotari, undertakes projects such as school building or road construction, financed by government but utilising local community labour. The government is trying to build on the corporate spirit engendered by social movements.

Indigenous groups have gained strength from cooperation, and not only in the defence of environmental issues. Driven by violence and oppression in rural areas, a number of Indigenous Guatemalan organisations came together to set up COMG (The Council of Maya organisation of Guatemala). This umbrella organisation, born originally in 1984 but not officially established until 1990, provided a platform where different movements could bring their demands and experiences together. COMG attempts to actively defend the right of Indigenous self-determination by encouraging member groups to initiate development projects that not only benefit the Maya in predictable ways, but also give them an awareness of their cultural and human rights. Successes have mostly occurred with small-scale local projects, but there has also been a recovery of Maya culture which goes well beyond the individual project level (Hertzog 1993).

Chilean organisations too benefited from the coordination at both the local and the national levels. The latter produced massive street demonstrations in 1983 which continued to harrow the dictatorial regime. Protesting at the economic conditions resulting from government policies as well as at the repressive measures of the regime, the demonstrations helped to undermine military rule.

As a result, popular social movements are often attributed with a role in the redemocratisation process that has taken place in Latin America. As mentioned in Chapter 5, the publicity for human rights gained by the *madres* played a part in the downfall of the Argentine military. The Brazilian social movements were partly responsible for the mobilisations in 1984 that eventually pushed the military into an acceptance of civilian rule.

However, even the major achievements of these movements have only been possible where other supporting factors were in place, and it may be the case that they have reached their peak. Hampered by the lack of any long-term clear ideology or organising principle, they can not easily take on state structures of power. Their only strength lies in numbers and the willingness to mobilise. Moreover, in forming larger networks and alliances they risk losing the internal democracy and personal involvement which have been their hallmarks.

CONCLUSION

The changes that have been taking place in Latin America, whether those resulting from economic growth or economic crisis, have not brought about any redistribution of income in favour of the low-income groups; the trend has been if anything towards a greater polarisation of wealth. Neither the dominant and bureaucratic–technical classes (due to their lack of motivation) nor the organised working class (due to its size and nature resulting from peripheral capitalist development) have been revolutionary forces in the region. Perhaps, as Becker has suggested for the bourgeoisie (Becker 1983) and Roxborough for the proletariat (Roxborough 1984), it is expecting too much to place the burden of revolution on the shoulders of any one class. The two major revolutions of modern times in the region – Cuba (1959) and Nicaragua (1979) – were both brought about by multi-class coalitions. The new social movements represent a multi-class force for change, but so far have in general been limited in their achievements.

FURTHER READING

Becker, D. (1983) *The new bourgeoisie and the limits of dependency: mining, class and power in 'revolutionary' Peru.* Princeton University Press, Princeton.

Eckstein, Susan (1989) *Power and popular protest, Latin American social movements*. Berkeley, London, University of California Press.

Escobar, A., Alvarez, S., (1992) *New social movements in Latin America, identity, strategy and democracy*. Westview Press, Boulder.

Felix, D. (1983) Income distribution and the quality of life in Latin America: patterns, trends and policy implications. *Latin American Research Review*, 18(2).

Foweraker, J., Craig, A. (1990) *Popular movements and political change in Mexico*, Lynne Reinner Publishers, Colorado and London.

Portes, A. (1985) Latin American class structures: their composition and change during the last decades. *Latin American Research Review*, 20(3).

Roxborough (1984) *Unions and politics in Mexico: the case of the automobile industry*. Cambridge University Press, Cambridge.

Chapter

9

The state

There are various reasons for the importance of the state in contemporary Latin America. Since industrialisation and economic growth have so often been perceived as the motor forces of development, one of the major concerns of development theory has been the source of capital necessary for these processes.

It is not just the bourgeoisie and multinationals who provide the investment for capital accumulation but also, in many countries, the state. This triple alliance of capital has become very important for economic growth in the more industrialised of the Latin American countries, with the state in some cases playing the dominant part. The state is able to manipulate such a role, for not only can it direct its own capital into the economic sectors where growth is required, but it can also influence its powerful partner, foreign capital, into the desired areas of the economy.

The state has emerged as such an important institution, not only because of its role in capital investment, but also as an employer. The expansion of state education and health facilities, as well as its economic activities and general growth of bureaucracy, mean that far more Latin Americans now work for the state. Their salaries, conditions of service and prospects for the future depend on the continuing prosperity of the state. The state bourgeoisie has expanded in numbers and increased its influence. Such was the expansion of the government sector in Ecuador, following the state's appropriation of oil revenues, that by 1982 employees of the state made up 22.9 per cent of the EAP (ISS–PREALC 1983).

It has been shown that one of the fundamental problems facing the poor is lack of access to secure well-paid work. As job opportunities are shaped by government policies, as well as educational opportunities, these are other areas in which the state has the power to make important contributions to development.

Is the state, then, becoming the new promoter of development? This chapter examines the explanations that have been put forward for the growth of the strong state and then looks at the role of the state towards development. It has been stressed that the 1960s and 1970s were periods of economic growth for Latin America, during which levels of growth at times surpassed those of the developed nations. In many Latin American countries this growth has been characterised by state-sponsored industrialisation. The stronger the state, the more able it is to influence the economy, and so the development of the authoritarian state has facilitated state sponsorship. In most, but not all, cases it is military governments that have been responsible for the authoritarian nature of the state.

Mexico has frequently been considered authoritarian, despite the existence of an election process to appoint the ruling party and the government of the day. The apparently elected ruling party has been the same one for about sixty years and has gained and maintained considerable control over the state apparatus in that time. Control is exercised through the integration of potential sources of opposition into the state apparatus, where they can be managed from above. The reciprocity which keeps this system going has already been described in the case of the labour movement. If for any reason these mechanisms fail to curb an insurrection, then the army can be called upon – as the events of 1968 showed. In that year, hundreds were killed by the army, after a massive student demonstration.

The argument of this chapter is that the policies of the authoritarian state have encouraged economic growth but at the same time exacerbated the accompanying social costs – a pattern of development which is characteristic of the region as a whole, but which is exaggerated by those republics with authoritarian states. This is partly due to monetarist policies, which authoritarian regimes are able to implement so fully because they can use force to put into practice the less palatable aspects. For society these policies have meant increased production and wealth and the development of a substantial, wealthy and European-style middle class, but they have also contributed to poverty, unemployment and underemployment and increasing inequality. The military regimes have now given way to democracies, but the legacy of the military has made development programmes difficult for the new governments. The new democracies have relied less on the state and more on market forces for the implementation of economic policies. Neither military regimes nor democracies have been very successful in reducing inequalities in the region.

THE AUTHORITARIAN STATE

Explanations for the growth of authoritarianism first focused on the military. In 1964 the military seized power in the largest country in Latin America, Brazil. It was a scenario which was soon to be repeated in perhaps the most advanced (in terms of economic development and indices of modernisation) Latin American country of the time, Argentina, and in one which had enjoyed an unprecedented period of constitutional governments, Bolivia. In 1968 Peru, too, experienced a military coup, and in 1973 the two countries in which democratic institutions were widely regarded as being securely planted, Chile and Uruguay, succumbed to the military. By that year, ten Latin American countries were governed by the armed forces. At first it was thought that the officers in power would adopt a 'caretaker' role, maintaining law and order for a short spell until the country was ready again for democratic rule, but contrary to these views the military stayed in power.

The first comprehensive theories of the military emphasised their increasing professionalisation and argued that military governments would become more infrequent with the expansion of the middle class (Lieuwen 1961; Johnson 1964). Johnson believed that military influence on politics could be beneficial to society because they were able to mobilise modernisation programmes, such as the colonisation of virgin territories and the building of roads. José Nun, however, having witnessed the military coups of the middle of the decade, saw the officers' regime as a more permanent feature of Latin American societies (Nun 1967). He distinguished between military intervention in the more backward and the more advanced republics, suggesting that in the latter it takes the form of a middle-class military coup. He argued that the army officers who came to power were allied to, and controlled by, the middle class. The latter were not strong enough on their own to fill the power vacuum left by the declining oligarchy but, through the armed forces, they could implement their policies. Unlike Johnson, Nun considered the middle class to be conservative, and so he characterised the military as being unfavourable to progress and social reforms.

The permanence of the armed forces in power required a rethinking of the role of the state. The move from democracy to authoritarianism did not fit the stages of modernisation theory, for it represented a move backwards to a more traditional order. On the other hand, some sociologists pointed out that these governments were embarking on modernising programmes. The Brazilians were in the process of developing Amazonia, by clearing the jungle and building settlements there, as well as encouraging highly technological and sophisticated industrial development. This led to the view that authoritarian regimes are not necessarily part of a traditional order, but have a part to play in the development of the LDCs (Malloy 1977).

O'Donnell was concerned to show that these regimes were the out-come of precisely those processes, such as the expansion of market rela-tions, industrialisation and increasing political participation, that modernisation theorists had identified as leading to democracy (O'Donnell 1973). But O'Donnell emphasised that these processes resulted in government policies which were not democratic, redistribu-tive or humanitarian, thus adding more evidence to disprove the modern-ists. O'Donnell's term 'bureaucratic–authoritarianism' was first used by him to refer to the state in Argentina and Brazil after the mid-1960s, but has later been used by other social scientists to refer to various military states.

The applications of authoritarianism to Latin America have been based on the original model developed by Linz to explain Franco's Spain (1970). Linz defines authoritarian regimes as 'political systems with limited, not responsible, political pluralism: without elaborate and guid-ing ideology (but with distinctive mentalities); without intensive and extensive political mobilization (except some points of their develop-ment); and in which a leader (or occasionally a small group) exercises power within formally ill-defined limits but actually quite predictable ones' (Linz 1970: 255).

The main features of the `bureaucratic–authoritarian' state are, firstly, a form of government which ensures the domination of society through a class structure by a bourgeoisie that is externally orientated. Institutionally, it is made up of a coalition of coercive and technocratic agencies and is technically rational. It involves a system of political exclu-sion which denies the mass of people the rights of citizenship. Moreover, this system encourages foreign capital and promotes high levels of social inequality (Philip 1985). The links with foreign capital, encouraged by the 'bureaucratic-authoritarian' state, gave O'Donnell some affinity with dependency theorists, though most of his other ideas have developed out-side the dependency framework.

O'Donnell's very influential ideas have aroused some criticisms. Munck argues that these states (and in particular Argentina) are not so much bureaucratic, in a Weberian sense, as technocratic. He adds that authoritarianism is too vague a word, especially as it really acts as a euphemism for repression (Munck 1985).

Others, theorising about the state, have developed the theme of authoritarianism, emphasising in particular the structures of corporatism that are so often an aspect of authoritarian regimes. This mode of organ-ising state and society has been aptly described by Malloy in the follow-ing way:

> ... that each of these regimes is characterised by strong and relatively autonomous governmental structures that seek to impose on the society a system of interest representation based on enforced limited pluralism. These regimes try to eliminate spontaneous interest articulation and establish a

limited number of authoritatively recognised groups that interact with the governmental apparatus in defined and regularised ways. Moreover, the recognised groups in this type of regime are organised in vertical functional categories rather than horizontal class categories and are obliged to interact with the state through the designated leaders of authoritatively sanctioned interest associations.

(Malloy 1977: 4)

Malloy suggests that corporatism has become a dominant feature of the contemporary Latin American state, especially in the more industrially advanced countries, and that modernising authoritarian regimes with corporatist structures are likely to remain part of political life.

Since the corporatism of Latin America is distinct from the forms of the state featured in the capitalist and socialist paths of development, it needs some explanation. It has been argued that this mode reflects a Hispanic–Catholic tradition which has always existed beneath the surface, while it has been fashionable for other constitutional forms to come and go. But this reasoning does not explain why corporatism should have come to the fore at this particular point in time, nor does it account for its orientation towards modernisation.

Malloy argues that two other processes have been crucial in the growth of corporatism: delayed dependent development and populism (Malloy 1977). His explanation for the growth of corporatism is very similar to O'Donnell's account of the factors underlying the development of the 'bureaucratic–authoritarian' state. According to Malloy, delayed dependent development brought about a general crisis of public authority, which democratic regimes have been unable to resolve. The crisis arose because modernisation was brought to Latin America from abroad and imposed on traditional structures of patronage and particularism. This created a multiplicity of societal interests which could not easily be integrated into one political structure with a guarantee of stability. The big difference between Malloy and O'Donnell is that while the former emphasises modernisation, O'Donnell stresses the social costs of modernisation.

Malloy's main argument, however, is that populism was the response to the first crisis of delayed dependent development, and that this in turn laid the foundations for corporatism. Since populism has been a prominent feature of Latin American political systems in certain historical periods, it is worth turning attention, briefly, to this political form.

Populism is a movement that is organised on the basis of community rather than class, with a charismatic figure as a leader. But because a movement claims to be classless it is not necessarily so and, indeed, in Latin America populism has usually taken the form of class alliances. Initially a response to the collapse of the export-orientated economy, populism briefly united the middle and working classes in order to fill the power vacuum that had developed.

The disruption of export-based economies, brought on by the depression of the 1930s and 1940s, led to a desire for autonomous national development and the implementation of import substitution. The economic crisis engendered the collapse of hegemonic control of the oligarchy, who lost the support of the middle class, some of whom were losing their jobs. Although sometimes internally divided, as in Argentina between the export bourgeoisie and the small industrialists, the bourgeoisie pushed to the front of the political struggle. In order to strengthen their weak position, they sought allies elsewhere, the obvious choice being the working class. Populism, by using ideology that offered rewards to the workers, was the way in which the bourgeoisie attempted to construct a coalition and win support of the working class. The aim was to create coalitions powerful enough to control the state, and then, having gained power, the bourgeoisie could assert itself and the subordinate position of the working class.

The populists saw the problems of Latin America as stemming from dualism, oligarchic control, dependence and a weak state. Thus they wanted economic independence, an end to the semi-feudal structures of the countryside and social justice for all sectors. The agency which was to try to achieve this was the state. Populism was therefore 'statist', but the state was to be controlled by a multi-class alliance. Although populism advocated a pluralist coalition to achieve reform, the bourgeoisie were to be the paternalistic leaders in this, offering education and improvements to their followings in return for their support. This uneven exchange was, clearly, clientelist rather than one reflecting an equal relationship between classes. The strong ties in populism were, therefore, vertical, between patron and clients, rather than horizontal and, because of the emphasis on the state, populism was based on an implicit corporatist image of socio-political organisation. With populism, the state actively brought the masses into politics, giving a minority certain benefits, as is characteristic of clientelism.

Populism had its achievements: the Peronist period in Argentina brought about very definite improvements for the working class; Brazilian workers gained better working conditions from Vargas; and APRA has remained a strong political party for a long time in Peru. In general, populism had by the 1950s weakened the power of the traditional élites, promoted import substitution (which increased the significance of the industrialists and organised labour) and stimulated a general increase in political mobilisation.

But, despite the expansion of the state apparatus by the populists, the state itself did not increase its power or efficiency, mainly because of the continuation of dependency. The economic crises of the 1960s showed that the problems had not been solved and dependency still continued, though now often in the form of multinationals. The bourgeoisie had not proved strong enough to fill the power vacuum left by the declining oligarchy. Radicalised by the Cuban revolution, some Latin American groups were using political protest to push for reform in society. Under the guise of nationalism and security, the military stepped into the vacuum, seizing control of the extensive state apparatus that the populists had previously set up.

The military's seizure of power represented a move from the potentially democratic corporatism of the populists to the authoritarian corporatism of the soldiers and technocrats (Malloy 1977). Populism had paved the way by setting up vertical corporatist structures, but the armed forces have adapted these to more authoritarian ends. The military have wanted to achieve state-sponsored industrial growth and, in order to attain this, they have had to increase the power and autonomy of the state and consciously to favour the groups they feel most likely to be successful at the expense of others. This has led to authoritarianism and repression. Unlike the populist period, the state in the 1960s became powerless to prevent benefits being drained off by multinationals. In order to gain a measure of control over both transnationals and society, the bourgeoisie have favoured policies that are incompatible with liberal democracy.

The corporatist nature of the Mexican state can be seen in the way that groups are linked to the governing party. Enforced limited pluralism exists in the form of the various influential groups, such as business associations, unions and peasant organisations, having vertical ties with the ruling party. The resulting network that links the president of Mexico with peasants and factory workers acts more as a form of control than a means of interest representation.

The idea that the state has increased its power and autonomy challenges the crude interpretation of Marx that the state is the tool of the ruling class. Poulantzas, taking a more rigorous Marxist stand, argues that relative autonomy of the state can strengthen capitalism and therefore is not necessarily bad news for the dominant class (Poulantzas 1975). He points out that the function of the state in a determinate social formation and the interests of the dominant class in this formation coincide because of the system itself. The direct participation of members of the ruling class in the state apparatus is not the cause but the effect. Thus, for Poulantzas, the social background of the ruling class is not really relevant because, whatever their background, members of the ruling class want to preserve the society in which they are successful. Capitalist society requires a bourgeoisie and a proletariat to function and, since the ruling class wishes to preserve and prosper capitalism, it has to protect the class relations of capitalism and must, therefore, ensure the continued supremacy of the bourgeois ruling group. Since the role of the state is to preserve the class system of capitalist society, the interests of the state and the ruling class are entwined and ensure the continuing power of the ruling class.

This argument leads Poulantzas to his concept of the relative autonomy of the state. He suggests that capitalist society might benefit from

a state bureaucracy whose members are not drawn from the same dominant social class. Poulantzas argues that a ruling group might suffer divisions according to the different interests of the factions within it. If the
members of the state apparatus are not drawn from the dominant social
class, they will not identify with one particular faction nor try to promote
its interests in opposition to another faction. Instead, standing outside
these differences, they will be able to see what actions would best serve
the development of capitalist society and, since they share interests with
the state, act accordingly. In this way, they can organise the hegemony of
the whole class more successfully.

THE SOCIAL COSTS OF AUTHORITARIANISM

The policies of authoritarian regimes have only served to intensify the
existing concentration of income in the region. The growing polarisation
of wealth has been vigorous in Brazil and Mexico. In Argentina it was
slight until the 1970s, when it was exacerbated by the decline in real
wages. The decline of real wages has had the same effect in Chile and
Uruguay in the 1970s (Felix 1983). In the 1960s Uruguay was one Latin
American country which did not display extremes of income concentration, but by the late 1970s this situation had changed. In 1968 the
wealthiest 5 per cent of households accounted for 17 per cent of income,
but by 1979 their share had almost doubled, expanding to 31 per cent
(Gillespie 1985). In Guatemala, which has experienced a succession of
military dictators since 1954, income concentration is extreme. In 1980,
5 per cent of the population received 59 per cent of the national income,
while the poorest 50 per cent received 7 per cent (Howes 1986). In Chile,
after a decade of military politics, the poorest 40 per cent of the population received about half as much of total income as they did before the
militarisation of politics. In 1968 their share of national income was 13.4
per cent, but by 1983 it had dropped to 6.4 per cent, while in that year
the wealthiest 10 per cent received over 50 per cent of national income.
This was partly due to falling real wages, which dropped by 15 per cent
in 1981 and by another 16 per cent in 1982, and unemployment (LAWR
5 August 1983). By the end of the military regime in 1989, about half the
Chilean population lived in poverty and 20–30 per cent in extreme
poverty (Angell and Pollack 1990).

Eckstein points out that despite Mexico's revolution, her income distribution is no more egalitarian than other Latin American countries, since
about 80 per cent of the population receive little more than 40 per cent of
the national income. More significantly, their share of national income did
not increase with the expansion of domestic production (Eckstein 1977).

One of the characteristics of monetarist policies, as practised by
the Latin American authoritarian regimes, has been to raise levels of

unemployment. In Chile, unemployment rose from 4.6 per cent in the early 1970s, before militarisation, to 24 per cent in 1983 (LAWR 5 August 1983). In Santiago it was estimated at more than 25 per cent (Grugel 1986).

One of the major problems to emerge in the 1980s was the debt crisis, arising from the heavy borrowing which had fuelled the state-sponsored growth of the 1970s. To try to tackle the issue, the Latin American debtors put forward proposals which included the separation of old from new debt, the restoration of US interest rates to historic levels, flexible negotiations and recognition of the need for development. The USA, prompted by fears of political destabilisation, responded with the Baker plan which, although it promised new loans in 1986–88 by multilateral financial institutions and commercial banks in return for some economic 'liberalisation', was clearly inadequate.

The real burden of the debt crisis falls on the mass of the poor. They are the people who ultimately suffer from the resulting increased unemployment, falling output, inflation and balance of payment crises. The solution is often seen in terms of approaching the IMF, whose prescriptions usually include stringent measures, such as setting higher prices for basic goods and services, controlling wages, cuts in government expenditure on public services, 'privatisation' of public-sector activities and encouragement for exports. These sorts of measures are frequently unacceptable to the LDCs, with the result that economic recession gets worse and the plight of the poor, who number over 40 million in a country like Brazil, is exacerbated.

The immense repayments that took place between 1981 and 1984 were obtained at the cost of great sacrifices in living standards for the workforce of most countries. According to the Inter-American Development Bank, the main burden during this period fell on wage-earners because of the increase in inflation and real depreciation of currencies. The largest cumulative declines in real wages occurred in Brazil and Mexico, which have the largest debts, the declines in Chile, Peru and Venezuela being somewhat smaller (IADB 1985). The resulting drop in economic activity during this period had repercussions throughout society, in particular unemployment rose and the informal sector expanded. Standards of living were further eroded by programmes of fiscal austerity, which reduced expenditure on public services. The comment of the Inter American Development Bank on this situation is worth noting:

> In view of the substantial losses in real income suffered by the region in recent years, and the evidence that a disproportionate part of these losses have been concentrated in the lower income strata, to the extent that real wage containment remains a necessary element of the adjustment process, mechanisms will have to be found to shift some of the burden to the higher income groups in the interests of social justice and domestic peace.
>
> (IADB 1985: 12–13)

The discontent aroused by the debt and general economic malaise seriously shook governments. In Chile, the financial collapse in 1983 compelled the military junta to start renegotiating the foreign debt with the banks and the IMF, and opened the door to opposition from sectors as varied as elements of the business community and the trade union movement. A number of protest movements arose, culminating in the first day of National Protest (in May 1983), which marked a real challenge to the regime. The military remained in power, though over a deeply-divided nation, until the elections of 1989. The early 1980s in Brazil were marked by unprecedented food riots, especially in the poorest parts of the country, where supermarkets and stores were ransacked by uncontrollable mobs.

A WEAKENING STATE?

After a lengthy period of military rule, twenty years in the case of Brazil, South American countries returned to civilian regimes. Ecuador was the first (1979), followed by Peru (1980), Bolivia (1982), Argentina (1983), Uruguay (1985) and Brazil (1985). Venezuela and Colombia have experienced continuous civilian rule since 1958. Paul Cammack optimistically saw this as marking the end of a fifty-year cycle of political instability and the beginning of a long-term period of democracy for South America (Cammack 1985). He does, however, point out that the type of democracy offered is élitist and limited in its aims. He suggests that the power vacuum which existed in the class structure since the 1930s was filled and that Latin America now has a civilian élite 'which is capable of ruling on behalf of the dominant classes with the consent of the majority, and of perpetuating its rule through the mechanism of competitive party politics' (Cammack 1985: 39). Cammack implies that the failure of the armed forces is the last of the unsuccessful attempts at development since the export crisis of the 1930s, and that it has ushered in a period of bourgeois hegemony. This hegemony is bolstered by the ability of the new élite to present left-wing moves for change as threats to law and order and, therefore, to thwart them.

Since most of the these democratic states have adopted neo-liberal policies, this has entailed a market philosophy which reduces the role of the state in the economy, and structural adjustment policies have limited the state's ability to extend public services. Does this then mean a weakening of the state?

Rosenthal argues that this has occurred in a number of ways. The financial restrictions, characteristic of the 1980s and 1990s, along with previous excesses in the spending by government bureaucracies, have greatly limited the public sector in most countries (Rosenthal 1989). The excessive debts of the region lowered the reputation of Latin American countries abroad, although some had been attempting to join the leading industrial nations in economic groupings. In many cases the paucity of the state's financial backing has limited its capacity to govern. This incapacity has led to a perception of the state as weak and unable to deliver, which has contributed to the growth of popular social movements. These in turn through an increasing number of successes, though often small in scope, have undermined the power of the state.

Traditionally in Mexico much of the state's power rested on clientelist relationships between the different levels of PRI (Institutional Revolutionary Party). The economic crisis of the 1980s seriously impaired the government's capacity to reward its traditional clients in the exchange that characterises political clientelism. With the declining efficacy of relationships with the state, the poor looked to the new social movements to solve their problems. In Mexico City efforts to secure land tenure and other urban services and facilities have traditionally been channelled through established patron–client relationships with the state. This gave PRI sole control up until the 1980s. Once CONAMUP had been set up, the inhabitants of the capital had a rival to turn to, and, given the government's straightened circumstances, one that was as effective. So concerned was the mayor of Mexico City that he set up new organisations to structure relationships between the people and himself. Even with this veneer of power the new committees failed because the real channels of communication were perceived to be the protest movements (Jimenez 1988).

The increasing loss of legitimacy of PRI has been expressed at the polls, culminating in the disputed election in 1988 of Salinas de Gotari to the presidency. Once in power, Salinas de Gotari sought to strengthen his own position by removing rival powerful figures such as Mexico's trade union boss, and that of PRI by revitalising the economy.

In Chile the transition to democracy did not take place until the 1990s, by which time neo-liberal policies were well established. As a result, Hojman argues that Aylwin's economic policies represent a fundamental continuity between the 1980s and the 1990s (Hojman 1990). Economic policy was seen to be successful as the previous four years had witnessed high levels of economic growth. The social costs, however, had been high with estimates of poverty in Chile in the 1980s varying between 40 per cent and 50 per cent. The new democratic regime, mindful of the stringent policies of the IMF was cautious in its approach to social issues, promising only to spend more on education rather than give any specific undertakings. It is significant, however, that in order to alleviate poverty, they undertook, even before taking office, to put into effect a moderate increase in the tax rate.

There is no doubt that there has been a rolling back of the state, but has this had the desired social and economic effects? We have seen that privatisations and cutbacks in public spending have not only appeased the IMF, but also won international acclaim and provided some financial backing for the state project. There is, however, evidence to suggest that the long-term benefits may be disappointing and that the rolling back may be important. Felix argues that long-term economic advantages may be thwarted because of the failure of macroeconomic stabilisation programmes, the prolonged decline in public investment and the fact that the belief that foreign capital inflows will follow the new economic policies may be misplaced. Although initially attracting foreign capital, the new policies now have to compete with demands for 'First World' money from Eastern Europe.(Felix 1992).

A free-market philosophy not only influences economic trends but also leads to an erosion of state action in favour of the underprivileged. While there may be an argument that the free market is the most efficient form of resource management, it clearly has no moral direction. The World Bank has identified continuing poverty as one of the greatest challenges to development policy, pointing out that to combat the problem successfully it must be tackled on two fronts: the promotion of the productive use of the labour of the poor and through the provision of basic services, such as health, education and nutrition to the poorest sector of the population. Far from the withdrawal of the state, this requires a dynamic role for the public sector. Taylor-Dormond's analysis of the Costa Rican state shows the significant impact that state action can have in reducing poverty and income inequality. His research reveals the direct effects of public subsidies, provided through the education, health, food, housing, water and social security programmes, on the incidence and intensity of poverty in Costa Rica. 'The effect of public social subsidies is to reduce total poverty from 26 per cent to 10 per cent and to narrow the global poverty gap by more than two-thirds' (Taylor-Dormond 1991: 131).

Moreover, political backing for privatisation is only fragile, with a number of social groups, such as the unions, the public-sector proletariat and some owners of local businesses who had supplied the domestic market and benefited from the protection offered by an interventionist state, actively opposing these moves. Opposition from these groups may help to bring about the return of more intervention in the public sector.

THE STATE AND INEQUALITY

Chapter 1 suggested that during both the years of economic growth and those of economic crisis inequalities have persisted and often become exacerbated. In most of Latin America the reliance on neo-liberalism in recent years has meant that the state has stood back from any involvement with income distribution, and allowed market forces to take their course. The character of economic development and these policies together have led to growing disparities in income levels.

The 1980s saw a general decline in the average material standard of living, resulting in a higher incidence of poverty and a widening of the gap between the upper and lower income groups. 'Income inequality in Latin America has been increasing as long as anyone can remember, but until the 1980s it did so with pauses and occasional reversals. . . . The Lost Decade breaks with this pro-cyclical pattern: it is the first extended depression in which income inequality has continued to worsen' (Felix 1992: 34). Income inequality reflects a growing social polarisation on a number of different scales.

Since the 1970s commentators have been noting the increasing difficulty in generalising about the region because of the growing differentiation between Latin American societies. Although there was considerable growth across the region in that decade, some countries, such as Argentina, Brazil, Mexico and Venezuela distinguished themselves by becoming the main recipients of international investment, thus falling into the category of NICs and separating themselves off from the rest. When it came to structural adjustment policies, some countries were able to adapt with greater success and so reinforced the already existing differences.

Regional disparities occur not just between countries but also between areas within countries. Brazil is a good example of this: the north-east has achieved an international reputation for poverty, while the south experienced the so-called economic miracle of the early 1970s. The compilers of the UN's Human Development Report have recognised the need for a disaggregated HDI profile, and by calculating the specific HDI for groups or regions within one country they can uncover startling differences in the national average. In Mexico, for example, the state of Oaxaca has an HDI 20 per cent lower than the national average.

Because more opportunities are to be found in the cities, as more jobs are created there through industrialisation, the disparity between urban and rural areas has grown. Despite capital-intensive development the jobs that have been created have been in urban areas. Industrialisation requires a more advanced infrastructure and also a support structure, thus creating a spin-off of wealth creation in the service sector. Although poverty has grown in urban areas in the last decade, acute poverty is still worst in rural areas. There 55 per cent of all the poor are indigent compared with around 35 per cent in urban areas (Feres and León 1990).

We have already seen in particular countries how the gap between upper- and lower-income groups expanded in the 1970s. Examining Latin America as a region, Weisskoff and Figueroa found two patterns of income redistribution during periods of economic growth. In the first there is a transfer of income shares from the poorest 90 per cent to the wealthiest 10 per cent, as in Brazil. In the second there is a twisting of the distribution away from the poorest 60 per cent and from the wealthiest 5 per cent toward a greater share for the middle groups, as in Mexico. What is evident from these findings is that either way, economic growth means a loss of relative shares to the poorest 60 per cent (Weisskoff and Figueroa 1976).

The 1980s crises hit everyone, but the poor particularly hard. Between 1980 and 1987 average income of wage/salary earners fell 15 per cent in the private formal sector and 30 per cent in the public sector. Concurrently, average real income fell 42 per cent in the urban informal sector, while the number of economically active persons in that sector rose 50 per cent (Feres and León 1990). Extremes can be seen in a country like Guatemala, where the wealthiest 20 per cent received 47 per cent of national income in 1970, 55 per cent in 1980 and 57 per cent in 1984, while the poorest 50 per cent had 24 per cent in 1970, 20 per cent in 1980 and 18 per cent in 1984 (Painter 1987). In an ECLAC study of ten Latin American countries (Argentina, Brazil, Colombia, Costa Rica, Guatemala, Mexico, Panama, Peru, Uruguay and Venezuela) the number of poor people rose by 28 million between 1980 and 1986 (Feres and León 1990). At the same time there were pockets of production modernisation, usually associated with the export of non-traditional goods, which created small oases of wealth accumulation. The situation for the 1980s can best be summed up in the words of Feres and León:

> The well-known and extensively analysed effects of the crisis, especially on the lower strata of the urban population, explain the fact that in the 1980s the drops in income and consumption were most marked in the lowest deciles of the income distribution scale, thus increasing the amount of poverty and the already high levels of inequality. Argentina and Uruguay, which had the relatively most equitable income distribution patterns of Latin America, suffered severe setbacks in this respect, sinking down closer to the level of countries with intermediate degrees of income concentration. Consequently, it is very likely that now, at the beginning of the 1990s, several countries of the region – specially those where their economic adjustments have meant pronounced drops in income – display greater inequity in distribution than around 1980 and similarly higher indexes of poverty, especially in urban areas.
>
> (Feres and León 1990: 151)

Various explanations for this increasing polarisation have been put forward. Booth's study of Central America shows that the development model based on the Alliance for Progress and applied in the 1960s and 1970s caused income and wealth to trickle up rather than down the social system. Central America received massive amounts of new investment which was largely taken up by import-dependent, capital-intensive manufacturing that consequently failed to absorb the rapidly growing labour supply. Economic growth followed, but there was no trickle-down effect of wealth through the expansion of employment and wages. Where policy-makers combined free-market capitalism with repressive labour policies, as in Nicaragua before 1979, Guatemala and El Salvador, wealth was redistributed in favour of the wealthy.

There were, however, variations within the region. Costa Rica and Honduras had mildly redistributive public policies, through income tax systems and the provision of health and education. These policies, along with the availability of agricultural land and employment, allowed some seepage of wealth to the working classes (Booth 1984). The development model is very similar to that experienced by most of Latin America and shows that where governments use repressive measures or where they do not intervene in the economic process, wealth moves away from the poor in both relative and absolute terms. This movement of wealth can only be altered by specific state policies of a redistributive kind.

The case for taxation is also made by James Painter in his study of Guatemala. He documents very thoroughly the gulf between the incomes of the rich and the poor, arguing that one of the main reasons for poverty is the very low level of taxes paid by the wealthy, who pay virtually less tax than any other élite in the Western hemisphere (Painter 1987). Many governments are reluctant to alienate their wealthy friends and political supporters by introducing higher taxes.

Although the state clearly has a role to play in income distribution, Felix suggests that consumption patterns in Latin American culture also contribute to socioeconomic polarisation. The wealthy have always expressed a preference for foreign-produced goods, which does little to promote domestic production with its potential for improved employment opportunities for Latin Americans (Felix 1983).

Societies which have not experienced growing social polarisation to the same extent, precisely because of their state policies, are the socialist states of Cuba and Nicaragua in the 1980s and also Costa Rica. Aspects of Cuba's social and economic development have been referred to elsewhere. With the crumbling of the socialist bloc in Europe, Cuba's future seems very uncertain and indeed some of the gains of the last three decades are already being eroded. Nicaragua's socialist period marked it off from the rest of Latin America during the 1980s, but current policies are bringing the country back into the fold. Costa Rica has pursued a path distinct from those of its capitalist and socialist neighbours. Though very clearly a capitalist state, it has over the years leaned towards higher taxation and more distributive policies than most Latin American countries and has not gone in for the excesses of authoritarianism as elsewhere.

Costa Ricans are very proud of their political stability and democracy which dates back to 1948. The reformist model which emerged then has dominated the state since and has been responsible for maintaining a strong tradition of welfare provision. As a result, social indicators of development are high for the region, and Costa Rica comes second to Cuba in terms of infant mortality, life expectancy, provision of health services and population coverage of social security (Mesa-Lago 1985). The state is able to finance this level of provision partly because of a progressive system of taxation which provides the wherewithal for income redistribution. It also has less of a drain on its resources than other states because of the lack of a national army – a feature for which Costa Rica has an international reputation. It should be noted that Costa Rica has had considerable support from the USA to maintain this political stability, particularly after the debt crisis when economic problems looked as if they would undermine all the gains of welfare provision. This has clearly increased the dependence of Costa Rica on the USA.

One might ask why a progressive taxation system is acceptable here but not elsewhere in Latin America, and why it is possible for this society to manage without an army in the centre of a turbulent region. One explanation is that, due to its historical background, the society has a more egalitarian social structure than most Latin American nations. Despite its name, Costa Rica was never very wealthy in the early years because of a lack of natural resources. There was not a sufficient surplus to support an élite, with the result that instead of a society of *hacendados* and *peons*, Costa Rica was made up mostly of yeoman farmers. Thus, when coffee cultivation (which did create wealth) began, a relatively egalitarian structure was already well entrenched. For much of recent history, land has been accessible for cultivation or work has been available either in coffee production or on the banana plantations, so that the problems of unemployment and underemployment have not been so apparent. Seligson also argues that the absence of a significant Indigenous population in the early years of colonisation meant that there was no distinct group of people which could be used as a subordinate labour force (Seligson 1984).

CONCLUSION

Latin America, in particular the more economically advanced states, has gone through a period of authoritarian regimes implementing policies of outward-orientated, capital-intensive development, followed by democracies pursuing neo-liberal ideas, influenced by structural adjustment policies. This has produced economic growth in the 1970s, a crisis in the 1980s and an economic recovery in the 1990s. Despite the cyclical nature of economic trends, many of the social indices of development have undergone a gradual decline over the decades. During the 1970s there was an improvement in the general standard of living, though this is disputed for the poor by some writers. Even given some absolute improvements, it is clear that the poor were not better off in relative terms as there was no fundamental redistribution of income. In the 1980s most of society suffered, with the poor bearing the brunt of the difficulties. The 1990s have produced an apparent economic flowering, though again, social indicators are not so positive. The incidence of poverty is still high:

there is no real change in income distribution; and a new factor – the spread of cholera (a disease associated with the poorest of living conditions) suggests that the economic revival has not affected some sectors of society.

The strong state has been shown to be a characteristic of the more industrially advanced Latin American countries. Military governments which were associated with this development have withdrawn from the political stage, leaving a legacy of economic malaise. The state has modernised in the sense of promoting economic growth, but the social costs of this policy have been great. Increasing concentration of income, rising unemployment, an expanding informal sector and falling real wages were compounded by the debt crisis. The policy of extensive borrowing to finance the economic growth has brought about an intolerable burden of repayments and debt-servicing. Democratisation was welcomed with considerable optimism in some quarters, but the tasks which the new democracies faced were formidable.

The growth of the strong state meant policies from above. The hallmark of the military was that they could impose their development plans, without any fears of democratic mechanisms preventing them being put into practice. This enabled them to implement the most stringent aspects of monetarist policy. The corporatist aspects of the state undermined the power of institutions such as trade unions to provide any meaningful opposition. Since governments gained control of organised groups, protest has occurred in the form of social movements. The incapacity of the state to provide adequate services, its economic failure and continual abuse of power have all contributed to the growth of these movements.

One of the characteristics of neo-liberalism is the withdrawal of the state from the economic arena. This reduced role for the state could be interpreted as a weakening of the institution. Moreover, the economic crises have severely impaired the capacity of governments to fulfil social welfare functions or carry out development promises. In a country like Mexico where the state has relied heavily on the reciprocity of political clientelism, the simple cooptation of contending forces has proved impossibly costly.

However, as Chapter 2 showed, this withdrawal through privatisation programmes has helped to fill the treasury coffers and so to make it possible for the state to contribute to debt repayment and to follow other policies. The state's role has changed to a less participatory one, though whether this will bring about economic and social development remains to be seen. What is clear is that the state welfare policies which have produced a measure of success in Costa Rica in the provision of health care and education and in the redistribution of income are not compatible with a neo-liberal philosophy, nor are they feasible under structural adjustment policies.

FURTHER READING

Angell, A., Pollack, B. (eds) (1993) *The legacy of dictatorship: political, economic and social change in Pinochet's Chile.* The Institute of Latin American Studies, University of Liverpool.

Malloy, J. (ed.) (1977) *Authoritarianism and corporatism in Latin America.* University of Pittsburgh Press, London.

O'Donnell, G. et al. (1986) *Transition from authoritarian rule.* Johns Hopkins University Press, Baltimore.

Ó'Maoláin, C. (1985) *Latin American political movements.* Longman, Harlow.

Philip, G. (1985) *The military in South American politics.* Croom Helm, Beckenham, Kent.

Remmer, K. (1991) *Military rule in Latin America.* Westview Press, Boulder.

Skidmore, T., Smith, P. (1989) (2nd edn) *Modern Latin America.* Oxford University Press, Oxford, New York.

Chapter

10

Conclusion

Contrary to much popular imagery, Latin America is a dynamic continent, where considerable social and economic change is taking place. This
book has examined how far these changes represent development for
Latin American society. The emphasis has been on social aspects, rather
than economic indicators, in an attempt to examine what benefits the
people of Latin America have gained from recent changes.

BROAD CHANGES

Although Latin America has been experiencing a variety of changes
across the region, certain broad themes emerge from these processes. On
the one hand, there is an increasing integration of society, as groups, such
as Indigenous communities, which were peripherally located, become
more involved with the market economy and the spread of mass media
incorporates a wider range of people into a universal value-system. On
the other hand, processes of change have brought about greater differentiation and produced a more complex social structure than existed
prior to this century. This differentiation can be seen at all levels and in
various distinct areas. It is a significant indicator that development theory
now offers no orthodox global approach which can be widely applied.
Instead, theory has become related to specific issues.

Differentiation has been documented in this book in a number of ways.
In economic terms there is a widening gap between the newly industrialising countries and the poorer republics of the region. In the countryside,

the traditional forms of labour have given way to increasing discrepancies in the size of smallholdings and a variety of forms of semi-proletarianisation and permanent agricultural labour, ranging from work for agribusiness to employment with well-to-do peasantry. The peasants have become internally differentiated by growing class affiliation. In the urban environment, the expansion of the informal sector has meant a rise in the forms of livelihood which involve varying degrees of irregularity and different types of links with the modern economy. In class terms, this has meant a more complex class structure, with the growth of the informal proletariat and informal petty bourgeoisie, but it has also brought about increasing income concentration.

Economic growth has taken place in the region, sometimes at very rapid rates, statistics revealing distinct levels of economic growth for the 1960s and 1970s. The economic debt problems of the early 1980s, and the resulting structural adjustment policies, however, resulted in the 'lost decade'. An alteration in the economic direction towards neo-liberalism suggests changing fortunes for the Latin American economies in the first half of the 1990s.

These economic trends have resulted in increased wealth for a minority of the population and an expanded middle class, with greater purchasing power, whose aspirations and role models are based on the middle class of the industrialised countries. There have also been improvements in some areas, such as nutritional levels (see Chapter 6) and the quality of life as expressed by physical indices (see Chapter 8) which affect a majority of the population. There has clearly been a shift from feudal to capitalist institutions, which can most clearly be seen in the agrarian structure, with the reduction of the number of *haciendas*. Agrarian reform has provided a more equitable distribution of landholding in some areas.

But the deepening of capitalism has also brought about the trend of social and economic differentiation. In many of the situations described in this book, a small minority have benefited, while the majority have suffered, thus increasing the difference in the levels of participation and widening even further the gap between the very rich and the very poor. Inequality (demonstrated by income distribution figures and class structure in Chapter 8) is softened, but also supported by the system of personalistic social relations (described in Chapter 5). The rural and urban poverty that this inequality implies is expounded in Chapters 6 and 7.

Linked to this is the trend towards more informal and, therefore, less secure forms of employment. In the countryside, semi-proletarianisation is on the increase and indeed a greater number of different types of semi-proletarianisation are appearing. In the Latin American towns, the informal sector has expanded in all countries. This means a flexible supply of labour for managers, but greater insecurity and low wages for workers. The insecurity lies in the lack of regular guaranteed work, the absence of benefits in the case of sickness and the mobility that may be necessary to move from job to job, which entails the disruption of supportive networks for the whole family.

The problems of underdevelopment have not been passively accepted by Latin American populations, but have frequently been met by a positive response from very different sectors of society. Indigenous groups threatened by the loss of their livelihoods have come together, often uniting tribes of a region for the first time to fight for their rights. Those such as the rubber-tappers, whose livelihoods are immediately threatened by ecological destruction have organised to protect the environment and their futures. Women are participating more in the public arena, not only in defence of issues associated with the private domain but also to work for neighbourhood improvements, human rights and gender-specific concerns.

It is notable that the Mexican earthquake of 1985 brought forth a very positive popular response, despite the vacillations of government which led to criticisms of inadequacy at the top. It was the popular organisation in four different sectors of Mexico City that made possible the creation of rescue teams, the speedy organisation of temporary shelters for the homeless and the development of programmes of mutual assistance between and within the different areas. The scale and seriousness of these groups impressed international bodies to such an extent that UNICEF channelled its campaign of preventative vaccination through a neighbourhood association.

PROBLEMS OF UNDERDEVELOPMENT

Chapter 1 posed the questions, what kind of economic development is necessary to enable welfare goals to be achieved, and can industrial growth be achieved without inequality? In order to attempt answers, the development process needs to be examined to see why and how problems have arisen.

One of the major sources of difficulty is that industrial growth has taken place on the basis of foreign capital. The great value of dependency theory lies in the way in which it pointed a finger at foreign capital and exposed the way that the centre had developed at the expense of the periphery. Others have argued, however, that industrialisation is necessary for development, capital is necessary for industrialisation and, therefore, foreign or multinational capital is the motor force of development. The linking of investment from outside Latin America with economic growth and the lack of welfare achievements can be seen in different areas.

One answer as to why economic growth has not reduced inequalities concerns the ways in which foreign capital, whether in the form of multinationals or loans, has contributed to polarisation. Firstly, while foreign capital has produced wealth, it has also brought about concentration (of wealth, income, productive forces, the benefits of development such as education, health facilities, etc.). The levels of capital accumulation and advanced technology that foreign capital makes possible act as a magnet for further investment and for people, at the same time producing the benefits in society. Secondly, much of the profit has gone abroad and still large sums leave the region to pay debts. Because of these two factors, the beneficiaries of development are few in number. Wealth has not trickled down to the mass of people because the types of employment available are structured towards the maximisation of profit and, therefore, where labour is plentiful, they represent cheap forms of labour, which offer little remuneration. Chapters 6 and 7 showed the ways in which the mass of people in rural and urban areas are limited in their opportunities. The strong state only encouraged these processes, as governments produced policies of development, financed by foreign capital and loans, which promoted economic growth with social costs. The neo-liberalism of the democratic regimes, though very different in its approach and apparently economically successful, has also done little to improve the lot of low-income groups or narrow the gap between rich and poor. With neo-liberalism's emphasis on market forces, the state is withdrawing from economic and welfare activities, thus exposing the poor to an economy often driven by international forces.

Inequalities have not been reduced, also because economic growth has not produced stable development. If the level of economic development reached at peak periods could be maintained, those who benefited would be able to consolidate their improvements and turn them into long-lasting achievements, which could be passed on to later generations as an acceptable level of prosperity. Stable development also means the growth of a class with sufficient savings and confidence in the economy to provide the local capital that is needed. The lack of internally generated development can frequently be traced to the international context. Examples in this book have demonstrated bursts of real economic development, which could not be maintained, at both micro- and macro-levels.

At the former level, sectors of the peasantry have at times been able to take advantage of market opportunities in international boom years and lost out in the downward fluctuations of the market. It was the global swings of the coffee market that meant a period of prosperity, only to be followed by poverty, the forced sale of land and migration for the Oaxacan peasantry. The Ecuadorian peasants benefited from the oil boom, in that the demand for food which this generated gave a boost to their production. But the ensuing economic crisis eroded their gains, demonstrating the flimsiness of development based on externally-generated growth. The oil boom and decline in general in Latin America has had a considerable effect upon employment. In the Mexican state of Tabasco, employment opportunities expanded from 1978–81 but, with the subsequent prices promoted by falling oil prices, many lost their jobs and had to migrate or return to subsistence agriculture to survive.

At national level, the Brazilian miracle, financed by multinational and foreign investment, produced very impressive growth rates in the late 1960s and early 1970s. The miracle came to an end because of rising oil prices and Brazil's inability to produce a cheap source of energy and because the repressive structure on which growth was built was vulnerable to any economic weaknesses. The motor force of change, foreign capital, generated its own problems in the form of massive debts.

This brings us on to another problem in financing development with borrowed capital – debt. The borrowing which countries such as Brazil, Mexico and Argentina relied on to finance the development process has brought almost as many problems as it sought to dispel. In Mexico, they say 'the devil gave us oil' to depict how the once-thought-of panacea has brought a multiplicity of problems. The changing world economy and increased interest rates have together made the large debts of the Latin American countries virtually unbearable. Latin American countries have managed to pay off some of the debts in stages, but the real burden of this has fallen on those who are least able to shoulder it. Privatisations of the early 1990s have helped to boost the incomes of Latin American states, but much of this new-found wealth has gone towards resolving the problems of the past rather than investing in the future.

At the same time development needs to take into account the environment. No real improvement can be said to be taking place if society's resource base is being destroyed. Quality of life in environmental terms, though difficult to measure, has now come to be accepted as a key concept.

PATHS OF DEVELOPMENT

The countries which have produced the highest rates of economic growth, but which have also experienced extremes in terms of social costs, are those characterised by growth from above. They have been governed by authoritarian regimes which have been in a position to impose their model of development. They are also areas where social movements, which epitomise change from below became active, which might suggest a growing impatience with the growth-from-above model. Growth from above has characterised socialist as well as capitalist societies, for Cuba, too, has a strong state and development follows a central plan. Does this mean that the focus of development for those involved should shift to the local level?

Because traditional mechanisms of political opposition were not available, the social movements have been articulated in new social spaces. Action does not emanate so much from political parties and unions, nor from the workplace but protest crystallises around neighbourhood problems, environmental concerns, Indigenous land rights and gender issues. Democratisation has not in general seen a dismantling of these movements, which in some areas have been greatly strengthened by the formation of umbrella organisations. It is notable that research in Chiapas in the 1970s and 1980s revealed the growth of independent peasant movements, inspired by claims for land rights, whose growing strength was due, amongst other things, to the manipulation of a network of alliances (Harvey 1990). By 1994 these movements had united with sufficient organisation and motivation to challenge the state. In most cases, however, movements have been issue-focused and aimed at problem resolution rather than confrontation. Perhaps development policies should be orientated more towards a basic-needs approach and look towards social movements for the identification of these needs, as indeed has already happened in Brazil, as quoted in the São Paulo Cost-of-Living-Movement example. These movements could act not only as sources of information, but also mechanisms of implementation as in the case of post-earthquake Mexico.

Alternatively, since industrialisation financed by foreign capital has not only created problems of underdevelopment, but also contributed to economic growth, the wealth produced by that growth could be used to greater mass advantage by radical, thorough and wide-ranging policies of redistribution. Gwynne has argued that the way forward for Latin America is further industrialisation, one of his points being that it is the most industrialised republics of the region that have best been able to confront the debt crisis (Gwynne 1985). It is, however, precisely these countries – Mexico and Brazil – where, as pointed out earlier, the poor have made the greatest sacrifices in order that repayments can be made. It is clear that not only have the poor not benefited from industrialisation, they are the ones paying the highest price to right the national-level problems that have arisen. If wealth does not automatically filter through to the rest of society, fundamental governmental redistribution programmes need to be implemented.

The radical and innovative redistribution programmes which have been tried, such as the Allende reforms in Chile and, at a different level, Mexico's attempt to utilise oil revenues through SAM, have often appeared attractive in theory but failed for a number of reasons. Redistribution programmes of this kind could be supported and aided by government policies of encouraging, through financial incentives, the growth of local-level, labour-intensive industries based on indigenous capital. If this were initiated on a small scale, sufficient investment might be found. The emphasis on labour-intensive activities would make a contribution towards easing the proliferation of irregular forms of labour.

Costa Rica has shown that the provision of welfare facilities, education and housing on a significant scale, the imposition of taxation and the redistribution of wealth are all possible within a framework of dependent capitalist development. The smaller gap between the rich and the poor than exists in the rest of Latin America, along with impressive welfare indices of development suggest a political way forward. However, not only has Costa Rica's very individual history given the country a unique profile, but the country's dependence on the USA, heightened by the debt crisis, makes this example difficult to follow and perhaps, for some, undesirable.

Whatever paths of development Latin American countries follow will depend upon the political complexion of their governments, international forces and the particular configuration of local, social and economic characteristics. One thing that is certain is that more research is wanted in the form of local-level studies of the day-to-day requirements of the needy.

Appendix

COUNTRY PROFILES

Argentina			
Area (km²)	2 776 656		
Total population (millions): 1992	33.1		
Urban population (percentage of total): 1990	86.3		
Average annual growth rate of total population (percentage): 1981–90	1.4		
Crude birth rate (births per 1000 population): 1990–95	20.3		
Crude death rate (deaths per 1000 population): 1990–95	9		
Infant mortality rate/crude infant death rate: 1990–95 (per 1000 population)	29		
Life expectancy at birth (years): 1990–95	71.4		
Male literacy (percentage): 1990	96		
Female literacy (percentage): 1990	95		
Percentage of households in conditions of poverty: 1970	8.0		
Composition of the labour-force (percentage): 1989–91	Agriculture 13	Industry 34	Services 53
Women in total labour-force (percentage): 1990	28.1		
GDP average annual growth (percentage)	1988 −2.7	1989 −3.8	1990 −1.1

Bolivia			
Area (km²)	1 098 581		
Total population (millions): 1992	7.83		
Urban population (percentage) of total): 1990	51.8		
Average annual growth rate of total population (percentage): 1981–90	2.8		
Crude birth rate (births per 1000 population): 1990–95	41.3		
Crude death rate (deaths per 1000 population): 1990–95	12		
Infant mortality rate/crude infant death rate (per 1000 population): 1990–95	93		
Life expectancy at birth (years): 1990–95	55.9		
Male literacy (percentage): 1990	85		
Female literacy (percentage): 1990	71		
Rural population below poverty line (percentage): 1977–87	85		
Composition of the labour-force (percentage): 1989–91	Agriculture 47	Industry 19	Services 34
Women in total labour-force (percentage): 1990	25.8		
GDP average annual growth (percentage)	1988 3.4	1989 3.5	1990 2.9

Brazil			
Area (km²)	8 511 965		
Total population (millions): 1992	156.28		
Urban population (percentage of total): 1990	74.9		
Average annual growth rate of total population (percentage): 1981–90	2.8		
Crude birth rate (births per 1000 population): 1990–95	26.1		
Crude death rate (deaths per 1000 population): 1990–95	8		
Infant mortality rate/crude infant death rate (per 1000 population): 1990–95	57		
Life expectancy at birth (years): 1990–95	66.3		
Male literacy (percentage): 1990	83		
Female literacy (percentage): 1990	80		
Percentage of households in condition of poverty: 1972	49.0		
Composition of the labour-force (percentage): 1989–91	Agriculture 28	Industry 25	Services 47
Women in total labour-force (percentage): 1990	27.4		
GDP average annual growth (percentage)	1988 0.0	1989 3.6	1990 –3.3

Chile

Area (km²)	756 629		
Total population (millions): 1992	13.60		
Urban population (percentage of total): 1990	85.9		
Average annual growth rate of total population (percentage): 1981–90	1.7		
Crude birth rate (births per 1000 population): 1990–95	22.5		
Crude death rate (deaths per 1000 population): 1990–95	6		
Infant mortality rate/crude infant death rate (per 1000 population): 1990–95	19		
Life expectancy at birth (years): 1990–95	72.1		
Male literacy (percentage): 1990	94		
Female literacy (percentage): 1990	93		
Population below poverty line (percentage): 1977–87	Urban: 27		
Composition of the labour-force (percentage): 1989–91	Agriculture 18	Industry 30	Services 52
Women in total labour-force (percentage): 1990	28.5		
GDP average annual growth (percentage)	1988 7.4	1989 9.9	1990 1.6

Colombia

Area (km²)	1 138 338		
Total population (millions): 1992	33.42		
Urban population (percentage of total): 1990	70		
Average annual growth rate of total population (percentage): 1981–90	2.1		
Crude birth rate (births per 1000 population): 1990–95	25.8		
Crude death rate (deaths per 1000 population): 1990–95	6		
Infant mortality rate/crude infant death rate (per 1000 population): 1990–95	37		
Life expectancy at birth (years): 1990–95	69.3		
Male literacy (percentage): 1990	88		
Female literacy (percentage): 1990	86		
Population below poverty line (percentage): 1977–87	Urban: 32		
Composition of the labour-force (percentage): 1980	Agriculture 34	Industry 24	Services 42
Women in total labour-force (percentage): 1990	21.9		
GDP average annual growth (percentage)	1988 3.6	1989 3.3	1990 3.5

Costa Rica			
Area (km^2)	50 000		
Total population (millions) (1992)	3.10		
Urban population (percentage of total): 1990	47.1		
Average annual growth rate of total population (percentage): 1981–90	2.8		
Crude birth rate (births per 1000 population): 1990–95	25.5		
Crude death rate (deaths per 1000 population): 1990–95	4		
Infant mortality rate/crude infant death rate (per 1000 population): 1990–95	17		
Life expectancy at birth (years): 1990–95	75.2		
Male literacy (percentage): 1990	93		
Female literacy (percentage): 1990	93		
Percentage of households in conditions of poverty (proportion of poor families): 1991[1]	9.8		
Composition of the labour-force (percentage): 1989–91	Agriculture 24	Industry 30	Services 46
Women in total labour-force (percentage): 1990	21.8		
GDP average annual growth (percentage)	1988 3.5	1989 5.7	1990 4.0

[1] Taylor-Dormond, M. The state of poverty in Costa Rica. *CEPAL Review* 43 (April 1991): 141.

Cuba			
Area (km²)	110 860		
Total population (millions): 1989	10.38		
Urban population (percentage of total): 1989	73		
Average growth rate of total population (percentage): 1975–85	1.32		
Crude birth rate (births per 1000 population): 1990	17.6		
Crude death rate (deaths per 1000 population): 1990	6.8		
Infant mortality rate/crude infant death rate (per 1000 population): 1988	13.3		
Life expectancy at birth (years): 1993	75		
Male literacy (percentage): 1990	95		
Female literacy (percentage): 1990	93		
Percentage of households in conditions of poverty	Not available		
Composition of the labour-force (percentage): 1986	Agriculture 17.6	Industry 22.2	Services 6.5[1]
Women in total labour-force (percentage): 1988	38.3		
GDP average annual growth (percentage)			

Source: Apart from the life expectancy, literacy, birth and death rate figures these statistics were taken from Stubbs, J. (1989) *Cuba: the test of time*, Latin American Bureau, London.
[1] Other: 53.7% (1986).

Dominican Republic			
Area (km²)	48 442		
Total population (millions): 1992	7.47		
Urban population (percentage of total): 1990	60.4		
Average annual growth rate of total population (percentage): 1981–90	2.3		
Crude birth rate (births per 1000 population): 1990–95	28.3		
Crude death rate (deaths per 1000 population): 1990–95	6		
Infant mortality rate/crude infant death rate (per 1000 population): 1990–95	57		
Life expectancy at birth (years): 1990–95	67.5		
Male literacy (percentage): 1990	85		
Female literacy (percentage): 1990	82		
Population below povery line (percentage): 1990	44		
Composition of the labour-force (percentage): 1989–91	Agriculture 46	Industry 15	Services 39
Women in total labour-force (percentage): 1990	15.0		
GDP average annual growth (percentage)	1988 1.3	1989 3.8	1990 −7.4

Ecuador			
Area (km^2)	270 670		
Total population (millions): 1992	10.74		
Urban population (percentage of total): 1990	56		
Average annual growth rate of total population (percentage): 1981–90	2.7		
Crude birth rate (births per 1000 population): 1990–95	30.9		
Crude death rate (deaths per 1000 population): 1990–95	7		
Infant mortality rate/crude infant death rate (per 1000 population): 1990–95	57		
Life expectancy at birth (years): 1990–95	66.6		
Male literacy (percentage): 1990	88		
Female literacy (percentage): 1990	84		
Population below poverty line (percentage): 1990	51		
Composition of the labour-force (percentage): 1989–91	Agriculture 30	Industry 24	Services 46
Women in total labour-force (percentage): 1990	19.3		
GDP average annual growth (percentage)	1988 11.4	1989 0.1	1990 1.3

El Salvador			
Area (km²)	20 935		
Total population (millions): 1992	5.40		
Urban population (percentage of total): 1990	44.4		
Average annual growth rate of total population (percentage): 1981–90	1.5		
Crude birth rate (births per 1000 population): 1990–95	36.0		
Crude death rate (deaths per 1000 population): 1990–95	7		
Infant mortality rate/crude infant death rate (per 1000 population): 1990–95	53		
Life expectancy at birth (years): 1990–95	66.5		
Male literacy (percentage): 1990	76		
Female literacy (percentage): 1990	70		
Population below poverty line (percentage): 1990	27		
Composition of the labour-force (percentage): 1980	Agriculture 43	Industry 19	Services 37
Women in total labour-force (percentage): 1990	25.1		
GDP average annual growth (percentage)	1988 3.1	1989 1.0	1990 2.5

Guatemala			
Area (km^2)	108 889		
Total population (millions): 1992	9.74		
Urban population (percentage of total): 1990	39.4		
Average annual growth rate of total population (percentage): (1981–90)	2.9		
Crude birth rate (births per 1000 population): 1990–95	38.7		
Crude death rate (deaths per 1000 population): 1990–95	8		
Infant mortality rate/crude infant death rate (per 1000 population): 1990–95	48		
Life expectancy at birth (years): 1990–95	64.8		
Male literacy (percentage): 1990	63		
Female literacy (percentage): 1990	47		
Population below poverty line (percentage): 1990	71		
Composition of the labour-force (percentage): 1989–91	Agriculture 48	Industry 23	Services 29
Women in total labour-force (percentage): 1990	16.4		
GDP average annual growth (percentage)	1988 4.3	1989 4.0	1990 3.9

Haiti

Area (km²)	27 750		
Total population (millions): 1992	6.76		
Urban population (percentage of total): 1990	28.3		
Average annual growth rate of total population (percentage): 1981–90	1.9		
Crude birth rate (births per 1000 population): 1990–95	35.3		
Crude death rate (deaths per 1000 population): 1990–95	12		
Infant mortality rate/crude infant death rate (per 1000 population): 1990–95	86		
Life expectancy at birth (years): 1990–95	56.7		
Male literacy (percentage): 1990	59		
Female literacy (percentage): 1990	47		
Population below poverty line (percentage): 1990	76		
Composition of the labour-force (percentage): 1989–91	Agriculture 50	Industry 6	Services 44
Women in total labour-force (percentage): 1990	41.6		
GDP average annual growth (percentage)	1988 −1.2	1989 −0.9	1990 −1.2

Honduras			
Area (km²)	112 088		
Total population (millions): 1992	5.46		
Urban population (percentage of total): 1990	43.7		
Average annual growth rate of total population (percentage): 1981–90	3.4		
Crude birth rate (births per 1000 population): 1990–95	37.1		
Crude death rate (deaths per 1000 population): 1990–95	7		
Infant mortality rate/crude infant death rate (per 1000 population): 1990–95	57		
Life expectancy at birth (years): 1990–95	65.8		
Male literacy (percentage): 1990	76		
Female literacy (percentage): 1990	71		
Population below poverty line (percentage): 1990	37		
Composition of the labour-force (percentage): 1989–91	Agriculture 36	Industry 17	Services 47
Women in total labour-force (percentage): 1990	18.8		
GDP average annual growth (percentage)	1988 5.0	1989 2.1	1990 –4.5

Mexico			
Area (km²)	1 967 183		
Total population (millions): 1992	89.54		
Urban population (percentage of total): 1990	72.6		
Average annual growth rate of total population (percentage): 1981–90	2.3		
Crude birth rate (births per 1000 population): 1990–95	26.6		
Crude death rate (deaths per 1000 population): 1990–95	5		
Infant mortality rate/crude infant death rate (per 1000 population): 1990–95	36		
Life expectancy at birth (years): 1990–95	70.4		
Male literacy (percentage): 1990	90		
Female literacy (percentage): 1990	85		
Percentage of households in conditions of poverty: 1967	34.0		
Composition of the labour-force (percentage): 1989–91	Agriculture 22	Industry 31	Services 47
Women in total labour-force (percentage): 1990	27.1		
GDP average annual growth (percentage)	1988 1.3	1989 3.1	1990 3.4

Nicaragua			
Area (km²)	139 000		
Total population (millions): 1992	4.13		
Urban population (percentage of total): 1990	59.8		
Average annual growth rate of total population (percentage): 1981–90	3.4		
Crude birth rate (births per 1000 population): 1990–95	38.7		
Crude death rate (deaths per 1000 population): 1990–95	7		
Infant mortality rate/crude infant death rate (per 1000 population): 1990–95	50		
Life expectancy at birth (years): 1990–95	66.3		
Total illiteracy (percentage): 1990–95	12.0		
Population below poverty line (percentage): 1990–95	Urban: 21	Rural: 19	
Composition of the labour-force (percentage): 1989–91	Agriculture 46	Industry 16	Services 38
Women in total labour-force (percentage): 1990	25.2		
GDP average annual growth (percentage)	1988 −9.3	1989 −2.5	1990 −2.0

Panama			
Area (km^2)	77 082		
Total population (millions): 1992	2.51		
Urban population (percentage of total): 1990	53.4		
Average annual growth rate of total population: 1981–90	2.1		
Crude birth rate (births per 1000 population): 1990–95	24.9		
Crude death rate (deaths per 1000 population): 1990–95	5		
Infant mortality rate/crude infant death rate (per 1000 population): 1990–95	21		
Life expectancy at birth (years): 1990–95	72.7		
Male literacy (percentage): 1990	88		
Female literacy: 1990	88		
Population below poverty line (percentage): 1990	25		
Composition of the labour-force (percentage): 1989–91	Agriculture 12	Industry 21	Services 67
Women in total labour-force (percentage): 1990	27.1		
GDP average annual growth (percentage)	1988 −15.7	1989 −0.9	1990 3.0

Paraguay

Area (km²)	406 752		
Total population (millions): 1992	4.52 (1992)		
Urban population (percentage of total): 1990	47.5		
Average annual growth rate of total population (percentage): 1981–90	3.1		
Crude birth rate (births per 1000 population): 1990–95	33.0		
Crude death rate (deaths per 1000 population): 1990–95	6		
Infant mortality rate/crude infant death rate (per 1000 population): 1990–95	39		
Life expectancy at birth (years): 1990–95	67.3		
Male literacy (percentage): 1990	92		
Female literacy (percentage): 1990	88		
Population below poverty line (percentage): 1990	35		
Composition of the labour-force (percentage): 1989–91	Agriculture 48	Industry 21	Services 31
Women in total labour-force (percentage): 1990	20.7		
GDP average annual growth (percentage)	1988 6.5	1989 6.1	1990 5.1

Peru			
Area (km^2)	1 280 219		
Total population (millions): 1992	22.45		
Urban population (percentage of total): 1990	70.2		
Average annual growth rate of total population (percentage): 1981–90	2.2		
Crude birth rate (births per 1000 population): 1990–95	29.0		
Crude death rate (deaths per 1000 population): 1990–95	8		
Infant mortality rate/crude infant death rate (per 1000 population): 1990–95	76		
Life expectancy at birth (years): 1990–95	64.6		
Male literacy (percentage): 1990	92		
Female literacy (percentage): 1990	79		
Population below poverty line (percentage): 1977–87	Urban: 49		
Composition of the labour-force (percentage): 1989–91	Agriculture 35	Industry 12	Services 53
Women in total labour-force (percentage): 1990	24.1		
GDP average annual growth (percentage)	1988 −8.0	1989 −11.4	1990 −4.3

Uruguay			
Area (km²)	176 215		
Total population (millions): 1992	3.13		
Urban population (percentage of total): 1990	85.5		
Average annual growth rate of total population (percentage): 1981–90	0.6		
Crude birth rate (births per 1000 population): 1990–95	17.1		
Crude death rate (deaths per 1000 population): 1990–95	10		
Infant mortality rate/crude infant death rate (per 1000 population): 1990–95	20		
Life expectancy at birth (years): 1990–95	72.5		
Male literacy (percentage): 1990	97		
Female literacy (percentage): 1990	96		
Population below poverty line (percentage): 1977–87	Urban: 22		
Composition of the labour-force (percentage): 1989–91	Agriculture 15	Industry 18	Services 67
Women in total labour-force (percentage): 1990	31.2		
GDP average annual growth (percentage)	1988 0.4	1989 1.3	1990 1.1

Venezuela			
Area (km²)	898 805		
Total population (millions): 1992	22.25		
Urban population (percentage of total): 1990	90.5		
Average annual growth rate of total population (percentage): 1981–90	2.8		
Crude birth rate (births per 1000 population): 1990–95	28.2		
Crude death rate (deaths per 1000 population): 1990–95	5		
Infant mortality rate/crude infant death rate (per 1000 population): 1990–95	33		
Life expectancy at birth (years): 1990–95	70.3		
Male literacy (percentage): 1990	87		
Female literacy (percentage): 1990	90		
Percentage of households in conditions of poverty: 1971	25.0		
Composition of the labour-force (percentage): 1989–91	Agriculture 12	Industry 32	Services 56
Women in total labour-force (percentage): 1990	27.6		
GDP average annual growth (percentage)	1988 5.8	1989 –8.3	1990 –1.2

Sources: Data are taken from *Monthly Bulletin of Statistics*, August 1993 UN, New York; *World Tables*, A World Bank Publication. 1991 edition; *Report on Economic and Social Progress in Latin America*, Inter American Development Bank. Washington DC 1991; *World Resources, 1992–93 A Guide to the Environment*. The World Resources Institute in Collaboration with the United Nations Environment programme, Oxford, 1993; *Third World Guide 93/94*, Instituto del Tercer Mundo Uruguay 1992; *Statistical Analysis of Latin America (SALA)* Wilkie, J. W. (ed.), Vol. 30, Part 1 1992, UCLA Centre Publications, CA. *UN Demographic Year Book*, 1992 UN, New York.

Glossary

boias-frías	Rural workers who live in urban areas and have to take their food with them and eat it cold.
cacique	Rural boss.
callampas	Chilean shanty towns.
castanha	A nut tree grown in North-East Brazil.
coca	Coca, a plant whose leaves when chewed act as a mild stimulant.
colono	Resident estate worker allotted a piece of land for subsistence farming.
compadrazgo	Ritual sponsorship of children by godparents, instituting a set of permanent, personal ties between parents, children and godparents.
compadres	Parents and godparents who have a special relationship.
cordón	Units of industrial worker organisations set up in Chile in the 1970s.
ejido	Unit of agrarian reform in Mexico.
empate	Resistance technique developed by the rubber-tappers of the Amazon.
encomienda	A mechanism introduced by the Spaniards, which flourished in the sixteenth and seventeenth centuries, to control and exact tribute and labour services from the indigenous population.
farinha	Flour.
favela	Brazilian shanty town.

favelado	Inhabitants of the *favelas*.
fazenda	Hacienda in Brazil.
garimpeiros	Brazilian gold prospectors.
hacienda	Large estate often producing for export, which represents a social as well as an economic system.
inquilinos	Service tenants in Chile.
ladino	*Mestizo* in Middle America.
latifundio	Large estate.
machismo	Cult of male dominance, derived from the word *macho*, meaning male.
madre	Mother.
maquiladoras	Assembly plants which import component parts and make them up for export.
marianismo	Cult of female subordination in general, but superiority in spiritual matters.
mestizo	Someone of mixed European and Indian stock.
mulato	Someone of mixed negro and European stock.
municipio	Municipality.
palanca	Intermediary between patron and client (literally a lever).
panalinha	Informal group made up of people linked by personal ties, but in different occupations.
tortilla	Mexican maize pancake.
villas miserias	Argentine shanty town.

Bibliography

Abel, C., Lewis, N. (eds) (1993) *Welfare, poverty and development in Latin America*. Macmillan, Basingstoke.

Adams, W. (1990) *Green development, environment and sustainability in the Third World*. Routledge, London.

Albornoz, O. (1993) *Education and society in Latin America*. Macmillan, Basingstoke and London.

Alderson-Smith, G. (1984) Confederations of households: extended domestic households in city and country. In **Long, N. and Roberts, B.** (eds) *Miners, peasants and entrepreneurs*. Cambridge University Press, Cambridge.

Angell, A., Pollack, B. (1990) The Chilean elections of 1989 and the politics of the transition to democracy. *Bulletin of Latin American Research*, 9(1).

Angell, A., Pollack, B. (eds) (1993) *The legacy of dictatorship: political, economic and social change in Pinochet's Chile*. The Institute of Latin American Studies, University of Liverpool, Liverpool.

Archer, D., Costello, P. (1990) *Literacy and power: the Latin American battleground*. Earthscan Publications, London.

Archetti, E., Cammack, P., Roberts, B. (1986) *Sociology of 'developing societies': Latin America*. Macmillan, London.

Armstrong, W., McGee, T. (1985) *Theatres of accumulation: studies in Asian and Latin American urbanization*. Methuen, London and New York.

Arriagada, I. (1990) Unequal participation by women in the working world. *CEPAL Review* **40**, April, ECLAC, Chile.

Bailey, F. (1971) The peasant view of the bad life. In **Shanin, T.** (ed.)
Peasants and peasant society. Penguin, Harmondsworth.
Baran, P. (1957) *The political economy of growth*. Monthly Review
Press, New York.
Barke, M., O'Hare, G. (1985) *The Third World: conceptual frameworks
in geography*. Oliver & Boyd, Edinburgh.
Barnes, J. (1993) Driving roads through land rights: the Colombian
Plan Pacífico. *The Ecologist* 24/4 August.
Barrett, M., Phillips, A. (1992) *Destabilising theory: contemporary
feminist debates*. Polity Press, Cambridge.
Barrios de Chungara, D., Viezzar, M. (1978) *Let me speak! Testimony
of Domitila, a woman of the Bolivian mines* (translated from Spanish by
V. Ortiz). Monthly Review Press, New York and London.
Baster, N. (ed.) (1972) *Measuring development: the role and adequacy of
development indicators*. Frank Cass, London.
Bauer, P. (1981) *Equality, the Third World and economic delusion*.
Weidenfeld & Nicolson, London.
Becker, D. (1983) *The new bourgeoisie and the limits of dependency:
mining, class and power in 'revolutionary' Peru*. Princeton University
Press, Princeton.
Belén, A., Bose, C. E. (1993) *Researching women in Latin America and
the Caribbean*. Westview Press, Oxford.
Beneria, L., Roldan, M. (1987) *The crossroads of class and gender:
industrial house work, sub-contracting and household dynamics in Mexico
City*. University of Chicago Press, Chicago.
Binder, L. (1986) The natural history of development theory.
Comparative Study of Society and History, 28(1).
Birbeck, C. (1979) Garbage, industry and the 'vultures' of Cali. In
Bromley, R., Gerry, C. (eds) *Casual work and poverty in the Third
World*. John Wiley, Chichester.
Blaikie, P. (1985) *The political economy of soil erosion in developing
economies*. Longman, Harlow.
Blauert, J., Guidi, M. (1992) Local initiatives in Southern Mexico. *The
Ecologist* 22/6 November/December.
Bonilla, F. (1970) Rio's favelas: the rural slum within the city. In
Mangin, W. (ed.) (1970) *Peasants in cities: readings in the anthropology
of urbanization*. Houghton Mifflin Co., Boston.
Booth, D. (1976) Cuba, colour and the revolution. *Science and Society*,
40(2).
Booth, D. (1985) Marxism and development sociology: interpreting the
impasse. *World Development*, 13(7).
Booth, J. (1984) 'Trickle up' income distribution and development in
Central America during the 1960s and 1970s. In **Seligson, M.** (ed.)
The gap between the rich and the poor. Westview Press, Colorado.
Bossen, L. (1984) *The redivision of labour: women and economic choice in*

four Guatemalan communities. State University of New York Press, Albany.

Bossert, T. (1980) The agrarian reform and peasant political consciousness in Chile. *Latin American Perspectives*, **27**, Vol. 7, No. 4.

Bourricaud, F. (1967) *Power and society in contemporary Peru* (translated from French by P. Stevenson). Faber & Faber, London.

Bourricaud, F. (1975) Indian, mestizo and cholo as symbols of the Peruvian system of stratification. In **Glazer, N. and Moynihan, D.** (eds) *Ethnicity: theory and experience.* MA, Harvard University Press, Cambridge.

Boyne, R., Rattansi, A. (eds) (1990) *Post modern society.* Macmillan, Basingstoke.

Brady, C. (1992) *How the development of Amazonia has affected the tribal Indians of the Grande Carajas programme.* Dissertation submitted for BA honours degree in Latin American studies, Portsmouth Polytechnic.

Brain, R. (1972) *Into the private environment: survival on the edge of our civilization.* George Phillip & Son, London.

Brandt, W. (1980) *North:South, a programme for survival.* Pan Books, London.

Brock, C., Lawlor, H. (eds) (1985) *Education in Latin America.* Croom Helm, London.

Bromley, R., Gerry, C. (eds) (1979) *Casual work and poverty in Third World cities.* John Wiley, Chichester.

Bronstein, A. (1982) *The triple struggle: Latin American peasant women.* WOW Campaigns Ltd, London.

Brookfield, H. (1975) *Interdependent development: perspectives on development.* Methuen, London.

Bunyard, P. (1989) Guardians of the Amazon. *New Scientist*, 16/12/89.

Burbach, R., Flynne, P. (1980) *Agribusiness in the Americas.* Monthly Review Press, New York and London.

Burgess, R. (1978) Petty commodity housing or dweller control? A critique of John Turner's views on housing policy. *World Development*, **6**(9/10).

Buvnic, M., Horenstein, N. (1986) *Women's issues in shelter, agriculture, training and institutional development.* USAID, Costa Rica.

Caipora Women's Group (CWG) (1993) *Women in Brazil.* LAB, London.

Cammack, P. (1985) Democratisation: a review of the issues. *Bulletin of Latin American Research*, **4**(2).

Cardoso, F. (1972) Dependency and development in Latin America. *New Left Review* **74** July/August.

Cardoso, F. (1973) Associated-dependent development: theoretical and practical implications. In **Stepan, A.** (ed.) *Authoritarian Brazil:*

origins, policies, and future. Yale University Press, New Haven.

Cardoso, F., Faletto, E. (1979) Dependency and development in Latin America. *New Left Review* 74, July/August.

Carrier, J. (1991) The crisis in Costa Rica: an ecological perspective. In **Goodman, D., and Redclift, R.** (eds) *Environment and development in Latin America,* Manchester University Press, Manchester.

Carrillo, J., Hernandez, A. (1985) *Mujeres fronterizas en la industria maquiladora.* Colección Frontera, Mexico, D.F.

Castells, M. (1982) Squatters and politics in Latin America: a comparative analysis of urban social movements in Chile, Peru and Mexico. In **Safa, H.** (1982) *Towards a political economy of urbanization in Third World countries.* Oxford University Press, Delhi.

Castillo, L., Lehmann, D. (1982) Chile's three agrarian reforms: the inheritors. *Bulletin of Latin American Research,* 1(2).

CEPAL Review 21 (1983), December.

Chaney, E. (1979) *Supermadre: women in politics in Latin America.* University of Texas Press for the Institute of Latin American Studies, Austin and London.

Chant, S. (1987) Family structure and female labour in Querétaro, Mexico. In **Momsen, J. and Townsend, J.** (eds) *Geography of gender in the Third World.* Hutchinson, London.

Chant, S. (1991) *Women and survival in Mexican cities.* Manchester University Press, Manchester and New York.

Chant, S. (1992) *Gender and migration in developing countries.* Belhaven Press, London.

Chapman, R. (1975) The first ten years. In **Doodsell, J.** (ed.) *Fidel Castro's personal revolution in Cuba: 1959–1973.* Alfred Knopf, New York.

Chilcote, R., Edelstein, J. (1974) *Latin America: the struggle with dependency and beyond.* John Wiley, New York.

Chinchilla, N. (1977) Industrialisation, monopoly capitalism and women's work in Guatemala. In **The Wellesley Editorial Committee** (eds) *Women and national development: the complexities of change.* University of Chicago Press, Chicago and London.

Clayton, E. (1983) *Agriculture, poverty and freedom in developing countries.* Macmillan, London and Basingstoke.

Cleary, D. (1991) The Greening of the Amazon. In **Goodman, D. and Redclift, M.** (1991) *Environment and development in Latin America: the politics of sustainability.* Manchester University Press, Manchester.

Cloke, P., Philo, C., Sadler, D. (1991) *Approaching human geography.* Paul Chapman Publishing, London.

Cockburn, A., Hecht, S. (1989) *The fate of the forest: developers, destroyers and defenders of the Amazon.* Verso, London.

Cockcroft, J. (1983) *Mexico: class formation, capital accumulation and the state.* Monthly Review Press, New York.

Collier, D. (1975) Squatter settlements and policy innovation in Peru. In **Lowenthal, A.** (ed.) *The Peruvian experiment: community and change under military rule.* Princeton University Press, Princeton and London.

Colman, D., Nixon F. (1986) (2nd edn) *Economics of change in less developed countries.* Allan Publishers, Oxford.

Commander, S., Peek, P. (1986) Oil exports, agrarian change and the rural labour process: the Ecuadorian Sierra in the 1970s. *World Development* 14(1).

Corkill, D. (1985) Democratic politics in Ecuador, 1979–1984. *Bulletin of Latin American Research* 4(2).

Corkill, D., Cubitt, D. (1988) *Ecuador: fragile democracy.* Latin America Bureau, London.

Cornelius, W. (1971) The political sociology of cityward migration in Latin America: toward empirical theory. In **Rabinovitz, F. and Trueblood, F.** (eds) *Latin America Urban Research* (1). Sage, Beverly Hills, California.

Craske, N. (1993) Women's political participation in Colonias Populares in Guadalajara, Mexico. In **Radcliffe, S. and Westwood, S.** (1993) *VIVA! Women and popular protest in Latin America.* Routledge, London and New York.

Cubitt, T., Cubitt, D. (1980) National minorities, cultural identity and class differentiation in Latin America: some methodological notes. *Journal of Latin American Studies* 1.

Cubitt, T. (1972) *The role of kinship and friendship links among Chilean industrialists.* Paper presented to the kinship and social networks seminar at the Institute of Latin American Studies, June.

Cubitt, T. (1983) Women in the Latin American labour force. *Journal of Area Studies* 7.

Curtis, R. (1983) *Two agrarian reforms: a case study of Nicaragua.* Dissertation submitted for BA honours degree in Latin American studies, Portsmouth Polytechnic.

Dankelman, I., Davidson, J. (1988) *Women and the environment in the Third World: alliance for the future.* Earthscan Publications Ltd, London.

Davies, R. (1979) Informal sector or subordinate mode of production? A model. In **Bromley, R. and Gerry, C.** (eds) *Casual work and poverty in Third World cities.* John Wiley, Chichester.

De Janvry, A. (1981) *The agrarian question and reformism in Latin America.* The Johns Hopkins University Press, Baltimore and London.

De Janvry, A., Ground, L. (1978) Types and consequences of land reform in Latin America. *Latin American Perspectives* 19(5).

Demographic Year Book (1991). The United Nations, New York.

de Soto, H. (1989) *The other path: the invisible revolution in the Third*

World (translated from Spanish by J. Abbott). Tauris, London.

Dore, E. (1988) *The Peruvian mining industry: growth, stagnation and crisis.* Westview Press, London.

Dore, E., Weeks, J. (1992) *The red and the black: the Sandinistas and the Nicaraguan revolution.* Institute of Latin American Studies Research Paper 28, University of London.

Dorey, E. (1990) *The effects of deforestation in the Brazilian Amazonia.* Dissertation submitted for BA honours degree in Latin American studies, Portsmouth Polytechnic.

Dos Santos, T. (1973) The crisis in development theory and the problem of dependence in Latin America. In **Bernstein, H.** (ed.) *Underdevelopment and development.* Penguin, Harmondsworth.

Durkheim, E. (1915) *The elementary forms of religious life* (translated from French by J. Swain). Allen & Unwin, London.

Eckstein, S. (1977) *The poverty of revolution: the state and the urban poor in Mexico.* Princeton University Press, Princeton.

Eckstein, S. (1989) *Power and popular protest: Latin American social movements.* University of California Press, Berkeley and London.

ECLA (1985) *Statistical Yearbook for 1984.* UN, Santiago de Chile.

ECLAC/FAO (1985) The agriculture of Latin America: changes, trends and outlines of strategy. *CEPAL Review* **27**, ECLAC, Chile.

Economist book of vital world statistics (1990). Hutchinson, London.

Ennew, J. (1989) *The next generation: lives of third world children.* Zed Books, London.

Escobar, A., Alvarez, S. (1992) *New social movements in Latin America: identity, strategy and democracy.* Westview Press, Boulder.

Evans, P. (1979) *Dependent development: the alliance of multinational, state and local capital in Brazil.* Princeton University Press, Princeton.

Fearnside, P. (1986) *The human carrying capacity of the Brazilian Amazon.* Colombia University Press, New York.

Felix, D. (1983) *Income distribution and the quality of life in Latin America: patterns, trends and policy implications.* Latin American Research Review 18(2).

Felix, D. (1992) Privatizing and rolling back the Latin American state. *CEPAL Review* **46**, April.

Feres, J., León, A. (1990) The magnitude of poverty in Latin America. *CEPAL Review* **41**, August.

Figueroa, M. (1985) Rural development and urban food programming. *CEPAL Review* **25**.

Filet-Abreu de Souza, J. (1980) Paid domestic service in Brazil. *Latin American Perspectives* 7(1).

Filguera, C. (1983) To educate or not to educate: is that the question? *CEPAL Review* **21**.

Fisher, J. (1993) *Out of the shadows: women's resistance in politics in Latin America*. LAB, London.

Forster, N., Handelman H. (1985). In Super, J. and Wright, T. (eds) *Food, politics and society in Latin America*. University of Nebraska Press, London.

Foster, G. (1967) *Tzintzuntzan: Mexican peasants in a changing world*. Little, Brown & Co., Boston.

Foster Carter, A. (1978) The modes of production controversy. *New Left Review* 107.

Foster Carter, A. (1985) *The sociology of development*. Causeway Press, Ormskirk.

Foweraker, J., Craig, A. (1990) *Popular movements and political change in Mexico*. L. Rienner Publishers, London and Colorado.

Frank, A. (1967) *Capitalism and underdevelopment in Latin America: historical studies of Chile and Brazil*. Monthly Review Press, New York.

Frank, A. (1969) *Sociology of development and underdevelopment*. Causeway Press, Ormskirk.

Freire, P. (1968) Educacão e conscientizacão. *Cuaderno* 25, CIDOC, Cuernavaca, Mexico.

Freyre, G. (1946) *The master and the slaves (Casa-Grande e Senzala): a study in the development of Brazilian civilization* (translated from Portuguese by S. Putnam). Alfred Knopf, New York.

Galjart, G. (1964) Class and 'following' in rural Brazil. *America Latina* 7(3).

García Guadilla, M.P. (1992) Symbolic effectiveness, social practices and strategies of the Venezuelan environmental movement: its impact on democracy. In Escobar, A. and Alvarez, S. (1992) *New social movements in Latin America: identity, strategy and democracy*. Westview Press, Boulder.

García Guadilla, M.P. (1993) Women, environment and politics in Venezuela. In Radcliffe, S. and Westwood, S. (1993) *VIVA! Women and popular protest in Latin America*. Routledge, London and New York.

George, S. (1976) *How the other half dies: the real reasons for world hunger*. Penguin, Harmondsworth.

George, S. (1990) *Ill fares the land: essays on food, hunger and power*. Penguin, Harmondsworth.

Germani, G. (1972) Stages of modernisation in Latin America. In Halper, S. and Sterling, J. (eds) *The dynamics of social change*. Allison & Busby, London.

Gerry, C. (1979) Small-scale manufacturing and repairs in Dakar: a survey of market relations within the urban economy. In Bromley, R. and Gerry, C. (eds) *Casual work and poverty in Third World cities*. John Wiley, Chichester.

Gilbert, A. (1974) *Latin America: a geographical perspective.* Penguin, Harmondsworth.

Gilbert, A. (1990) Latin America. Routledge, London.

Gilbert, A. (1994) *The Latin American city.* Latin America Bureau, London.

Gilbert, A., Gugler, J. (1992) (2nd edn) *Cities, poverty and development: urbanization in the Third World.* Oxford University Press, Oxford and New York.

Gilbert, A., Ward, P. (1985) *Housing, the state and the poor.* Cambridge University Press, Cambridge.

Gillespie, C. (1985) Uruguay's return to democracy. *Bulletin of Latin American Research* 4(3).

Global Outlook 2000 (1990): An economic, social and environmental perspective. United Nations Publications, New York.

Gmelch, G., Zenner, W. (1988) (eds) *Urban life: readings in urban anthropology.* Waveland Press, Prospect Heights.

Goldthorpe, J. (1975) *The sociology of the Third World: disparity and involvement.* Cambridge University Press, Cambridge.

Gonzalez Cassanova, P. (1968) *Dynamics of the class structure.* In Kahl, J. (ed.) *Comparative perspectives on stratification.* Little Brown & Co., Boston, Mexico, UK and Japan.

Goode, W. (1963) *World revolution and family patterns.* Free Press, Glencoe, Illinois.

Goodman, D., Hall, A. (1990) *The future of Amazonia: destruction or sustainable development?* Macmillan, London.

Goodman, D., Redclift, M. (1981) *From peasant to proletarian: capitalist development and agrarian transitions.* Basil Blackwell, Oxford.

Goodman, D., Redclift, M. (1991) *Environment and development in Latin America: the politics of sustainability.* Manchester University Press, Manchester.

Gracierena, J., Franco, R. (1978) Social formations and power structures in Latin America. *Current Sociology* 26(1).

Green, D. (1991) *Faces of Latin America.* LAB, London.

Griffin, K. (1981) Economic development in a changing world. *World Development* 9(3).

Grugel, J. (1986) *Some reflections on 'successful' authoritarianism: the Chilean phenomena.* Paper presented to the annual conference of the Society for Latin American Studies, Liverpool.

Guimarães, R. (1989) The ecopolitics of development in Brazil. *CEPAL Review* 38, ECLAC, Chile.

Gwynne, R. (1985) *Industrialization and urbanization in Latin America.* Croom Helm, London.

Halper, S., Sterling, J. (1972) (eds) *Latin America: the dynamics of social change.* Allison & Busby, London.

Hamilton, H. (1982) *The limits of state autonomy: post-revolutionary Mexico*. Princeton University Press, Princeton.

Hansen, N. (1981) Development from above: the centre down development paradigm. In **Stohr, W.** and **Taylor, D.** (eds) (1981) *Development from above or below? The dialectics of regional planning in developing countries*. John Wiley, Chichester.

Hardin, G. (1968) The tragedy of the Commons. *Science* 162, pp. 1243–8.

Hart, K. (1973) Informal income opportunities and urban employment in Ghana. *Journal of Modern African Studies* 11.

Harvey, N. (1983) *Mexico stirred not shaken: the problem of alcoholism*. Dissertation submitted for BA honours degree in Latin American studies, Portsmouth Polytechnic.

Harvey, N. (1990) Peasant strategies and corporatism in Chiapas. In **Foweraker, J., Craig, A.** (eds) *Popular movements and political change in Mexico*. L. Rienner Publishers, London and Colorado.

Hecht, S. (1985) Environment, development and politics: capital accumulation and the livestock sector in Eastern Amazonia. *World Development* 13(6): 663–84.

Hellman, J. (1986) Migration and urbanization. *Latin American Research Review* 21(1).

Henfrey, C. (1984) Between populism and Leninism – the Grenadian experience. *Latin American Perspectives* 42, Vol. 11, No. 3.

Hertzog, J. (1993) *Five hundred years of Indigenous resistance in Guatemala: the struggle for equality and dignity in 1992*. Dissertation submitted for BA honours degree in Latin American studies, Portsmouth University.

Hewitt de Alcántara, C. (1984) *Anthropological perspectives on rural Mexico*. Routledge & Kegan Paul, London.

Hojman, D. (1990) Chile after Pinochet: Alwyn's Christian Democrat economic policies for the 1990s. *Bulletin of Latin American Research*, 9(1).

Hooks, B. (1982) *Ain't I a woman?* Pluto, London.

Hopkins, T., Wallerstein, I. (eds) (1982) *World systems analysis: theory and methodology*. Sage Publications, Beverley Hills, California.

Howes, M.C. (1986) *Guatemala: the rise of the military and the growing response of the Catholic church*. Dissertation submitted for BA honours degree in Latin American studies, Portsmouth Polytechnic.

Huberman, L., Sweezy, P. (1969) *Socialism in Cuba*. Monthly Review Press, New York.

Huizer, G. (1973) *Peasant rebellion in Latin America*. Penguin, Harmondsworth.

Hull, G., Scott, P., Smith, B. (1982) *All the women are white, all the Blacks are men, but some of us are brave*. Feminist Press, New York.

Human Development Report (HDR) (1991, 1992 and 1993) United

Nations Development Programme, Oxford University Press, Oxford.

Humphrey, J. (1982) *Capitalist control and workers' struggle in the Brazilian auto-industry.* Princeton University Press, Princeton.

Hurtado, M. (1986) Teeming cities: the challenge of the urban poor. *Latin America and Caribbean Review.* World of Information, Saffron Walden.

IBT (1985) Brazil, Brazil. London.

ILO (1976) *Employment, growth and basic needs: a one world problem.* Geneva.

Imaz, J. (1964) *Los que mandan.* Editorial Universitaria, Buenos Aires.

International Labour Review (1961) July–August 1961.

Inter-American Development Bank (IADB) (1985) *Economic and social progress in Latin America: external debt: crisis and adjustment.* Washington DC.

Inter American Development Bank (IADB).(1991) *Economic and social progress in Latin America.* Johns Hopkins University Press for the IDB, Washington DC.

International Marketing and Data Statistics (1990) (16th edn). Euromonitor, London.

ISS-PREALC (1983) *Planning for basic needs in Latin America. Urban poverty, access to basic services and public places: a case study of the marginal suburbs of Guayaquil, Ecuador.* Working paper No. 7, Quito.

Iturralde, M. (1987) Evaluation of the strategy for health and population development in Latin America and the Caribbean. *Latin American Report* 3(2).

Jaquette, J. (1991) *The woman's movement in Latin America: feminism and the transition to democracy.* Unwin Hyman, London and Boston.

Jelin, E. (1990) *Women and social change in Latin America* (translated by J. Zammitt and M. Thompson). UN Research Institute for Social Development, Zed Books, London.

Jelin, E. (1991) *Family, household and gender relations in Latin America.* Kegan Paul/UNESCO, London.

Jimenez, E. (1988) New forms of community participation in DF: success or failure? *Bulletin of Latin American Research* 7(1).

Johnson, D. (1972) The national and progressive bourgeoisie in Chile. In Cockroft, J., Frank, A., Johnson, D. (eds) *Dependence and underdevelopment: Latin America's political economy.* Anchor Books, New York.

Johnson, H. (1992) Women's empowerment and public action. In Wuyts, M., Mackintosh, M., Hewitt, T. (eds) *Development policy and public action.* Oxford University Press, Oxford.

Johnson, J. (1958) *Political change in Latin America: the emergence of the middle sectors.* Stanford University Press, Stanford.

Johnson, J. (1964) *The military and society in Latin America.* Stanford University Press, Stanford.

Kahl, J. (1970) *The measurement of modernism.* University of Texas Press, Austin and London.

Kay, C. (1978) Agrarian reform and the class struggle in Chile. *Latin American Perspectives* 18, Vol. 3, No. 11.

Kay, C. (1989) *Latin American theories of development and underdevelopment.* Routledge, London.

Kendall, S. (1992) Indians seek a united voice. *Financial Times* December, 11.

King, J. (1990) *Magical reels: a history of cinema in Latin America.* Verso, London.

King, K. (1979) Petty production in Nairobi: the social context of skill acquisition and occupational differentiation. In **Bromley, R. and Gerry, C.** (eds) *Casual work and poverty in Third World cities.* John Wiley, Chichester.

Kitching, G.N. (1989) (revised edn) *Development and underdevelopment in historical perspective: populism, nationalism and industrialisation.* Routledge, London.

Laclau, E. (1971) Feudalism and capitalism in Latin America. *New Left Review* 67.

Lall, S. (1975) Is dependence a useful concept in analysing underdevelopment? *World Development* 3(11).

Lambert, J. (1967) *Latin America: social structure and political institutions* (translated from French by H. Katel), University of California Press; Cambridge University Press, London.

Latin America Bureau (1982) *Brazil, state and struggle.* LAB, London.

Latin American and Caribbean Women's Collective (1977) *Slaves of slaves: the challenge of Latin American women.* Zed Press, London.

LAWR (Latin American Weekly Report), 5 August 1983.

Leeds, A. (1969) The significant variables determining the character of squatter settlements. *América Latina* 12, (July–Sept).

Leeds, A. (1974) Brazilian careers and social structure: a case history and model. In **Heath, D.** (ed.) *Contemporary cultures and societies of Latin America.* Random House, New York.

Lehmann, D. (1982) Agrarian structure, migration and the state in Cuba. In **Peek, P. and Standing, G.** (eds) *State policies and migration.* Croom Helm, London.

Lenin, V. (1970) Imperialism, the highest stage of capitalism. In **Lenin, V.** *Selected works,* Vol. 1. Progress Publishers, Moscow.

Levine, D.H. (1981) *Religion and politics in Latin America: the Catholic church in Venezuela and Colombia.* Princeton University Press, Princeton; Guildford, Surrey.

Lewis, O. (1966) The culture of poverty. *Scientific American* 215(4).

Lewis, S.A. (1992) Banana bonanza: multinational fruit companies in Costa Rica. *The Ecologist* 22(6), November/December: 289–90.

Lieuwen, E. (1961) *Arms and politics in Latin America.* Praeger, New York.

Linz, J. (1970) An authoritarian regime: Spain. In **Allardt, E. and**

Rokkan, S. (eds) *Mass politics: studies in political sociology.* Free Press, New York and London.

Lipman, A. (1969) *The Colombian entrepreneur in Bogotá.* University of Miami Press, Coral Gables, Florida.

Lipton, M. (1977) *Why poor people stay poor: a study of urban bias in world development.* Temple Smith, London.

Lloyd, P. (1982) *A Third World proletariat?* Allen & Unwin, London.

Lobo, S. (1982) *A house of my own: social organisation in the squatter settlements of Lima, Peru.* University of Arizona Press, Tucson.

Logan, K. (1988) Latin American urban mobilisation: women's participation and self empowerment. In **Gmelch, G. and Zenner, W.** (eds) *Urban life: readings in urban anthropology.* Waveland Press, Prospect Heights.

Lomnitz, L. (1977) *Networks and marginality: life in a Mexican shanty town* (translated from Spanish by C. Lomnitz). Academic Press, New York and London.

Lomnitz, L. (1982) Horizontal and vertical relations and the social structure of urban Mexico. *Latin American Research Review* 17(2).

Long, N. (1977) *An introduction to the sociology of rural development.* Tavistock, London.

López Cordovez, (1982) Trends and recent changes in the Latin American food and agricultural situation. *CEPAL Review* 16.

López, C., Pollack, M. (1989) The incorporation of women in development policies. *CEPAL Review* 39, December. ECLAC, Chile.

MacDonald, N. (1991) *Brazil: A mask called progress.* Oxfam, Oxford.

MacEwen Scott, A. (1979) Who are the self-employed? In **Bromley, R. and Gerry, C.** (eds) *Casual work and poverty in Third World cities.* John Wiley, Chichester.

Mahoney, R. (1992) Debt for nature swaps: who really benefits? *The Ecologist* 22/3, May/June.

Mainwaring, S. (1987) Urban popular movements and democratisation in Brazil. *Comparative Political Studies* 20(2).

Malloy, J. (ed.) (1977) *Authoritarianism and corporatism in Latin America.* University of Pittsburgh Press, London.

Mangin, W. (ed.) (1970) *Peasants in cities: readings in the anthropology of urbanization.* Houghton Mifflin Co., Boston.

Martin, D. (1990) *Tongues of fire: the explosion of protestantism in Latin America.* Basil Blackwell, Oxford.

Martin, G. (1986) Carrying the can: Bolivian labour and the international tin crisis. *Journal of Area Studies* 13.

Martinez, J., Valenzuela, E. (1986) Chilean youth and social exclusion. *CEPAL Review* 29, Santiago.

Mattelart, A. (1978) The nature of communications practice in a dependent society. *Latin American Perspectives* 16, Vol. 5, No. 1.

Mendes, C. (1992) Excerpts from Chico Mendes' *Fight for the forest.*
Latin American Perspectives 72, Vol. 19, No. 1, Winter: 144–7.
Mendes, C., Gross, T. (1989) *Fight for the forest: Chico Mendes in his
own words.* LAB, London.
Mendes, R.B. (1987) The rural sector in the socio-economic context of
Brazil. *CEPAL Review* 33, December, ECLAC Chile.
Mesa-Lago, C. (1986) Social security and development in Latin
America. *CEPAL Review* 28.
Miller, M. (1991) *Debt and the environment: converging crisis.* UN
Publications, New York.
Miró, C., Rodríguez, D. (1982) Capitalism and population in Latin
American agriculture. Recent trends and problems. *CEPAL Review*
16.
Mitchell, S. (ed.) (1981) *The logic of poverty: the case of the Brazilian
North East.* Routledge & Kegan Paul, London.
Molyneux, M. (1985) Mobilization without emancipation? women's
interests, state and revolution in Nicaragua. In **Slater, D.** (ed.) *New
social movements and the state in Latin America.* CEDLA, Amsterdam.
Momsen, J.H. (1991) *Women and development in the Third World.*
Routledge, London and New York.
Montgomery, T. (1983) The church in the Salvadorean revolution.
Latin American Perspectives 36, Vol. 10, No. 1.
Morse, R. (1971) Trends and issues in Latin American urban research
1965–70. *Latin American Research Review* 6(1).
Moser, C. (1978) Informal sector or petty commodity production:
dualism or dependence in urban development? *World Development* 6.
Moser, C. (1989) Gender planning in the Third World: meeting
practical and strategic gender needs. *World Development* 17(11).
Munck, R. (1985) The 'modern' military dictatorship in Latin America:
the case of Argentina (1976–1982). *Latin American Perspectives* 47,
Vol. 12, No. 4.
Muñoz, H. (1980) *From dependency to development: strategies to
overcome underdevelopment and inequality.* Westview Press, Boulder,
Colorado.
Muñoz, H. (1981) The strategic dependency of the centres and the
economic importance of the Latin American periphery. *Latin
American Research Review* 16(3).
Murdoch, W. (1982) *The poverty of nations.* Johns Hopkins University
Press, Baltimore and London.
Myers, N. (1981) The Hamburger connection: how Central America's
forests became North America's hamburgers. *Ambio* 10(1): 3–8.
Nash, J. (1977) Women in development: dependency and exploitation.
Development and Change 8.
Nash, J., Safa, H. (eds) (1976) *Sex and class in Latin America.* Praeger,
New York and London.

Nations, J.D., Kormer, D.I. (1987) Rainforests and the Hamburger Society. *The Ecologist* 17(4/5): 161–7.

Navarro, V. (1976) *Medicine under capitalism.* Croom Helm, London and New York.

Nelson, C. (1992) *Tropical forest conservation and indigenous peoples in Ecuador's 'Oriente' region.* Dissertation submitted for BA honours degree in Latin American studies, Portsmouth Polytechnic.

Nelson, J. (1969) *Migrants, urban poverty and instability in developing nations.* Harvard University Center for International Affairs, Cambridge, Mass.

Nugent, S. (1991) The limitations of environment 'management' forest utilization in the Lower Amazon. In **Goodwin, D., and Redclift, M.** (eds) *Environment and development in Latin America: the politics of sustainability.* Manchester University Press, Manchester.

Nun, J. (1967) The middle class military coup. In **Veliz, C.** (ed.) *The politics of conformity in Latin America.* Stanford University Press, Stanford.

O'Brien, P. (1975) *A critique of Latin American theories of dependency.* In **Oxaal, I., Barnett, T. and Booth, D.** (1975) *Beyond the sociology of development.* Routledge & Kegan Paul, London.

O'Brien, P., Cammack, P. (eds) (1985) *Generals in retreat: the crisis of military rule in Latin America.* Manchester University Press, Manchester.

O'Donnell, G. (1973) *Modernization and bureaucratic authoritarianism: studies in South American politics.* University of California Press, Berkeley.

O'Donnell, G. et al (1986) *Transition from authoritarian rule: tentative conclusions about uncertain democracies.* Johns Hopkins University Press, Baltimore.

Ó'Maoláin, C. (1985) *Latin American political movements.* Longman, Harlow.

O'Riordan, T. (1981) *Environmentalism.* Pion, London.

Ortega, E. (1982) Peasant agriculture in Latin America: situations and trends. *CEPAL Review* 16.

Oxaal, I., Barnett, T., Booth, D. (1975) *Beyond the sociology of development: economy and society in Latin America and Africa.* Routledge & Kegan Paul, London.

Painter, J. (1987) *Guatemala: false hope, false freedom.* CIIR/LAB, London.

Palma, G. (1981) Dependency: a formal theory of underdevelopment or a methodology of analysis. In **Streeton, P. and Jolly, R.** (eds) *Recent issues in world development: a collection of survey articles.* Pergamon Press, Oxford.

Parkinson, F. (1984) Latin America, her newly industrialising countries and the new international economic order. *Journal of Latin American Studies* 16, Part 1.

Payne, J. (1965) *Labour and politics in Peru.* Yale University Press, New Haven and London.

Pearce, D., Markandya, A., Barbier, E. (1989) *Blue print for a green economy.* Earthscan, London.

Pearce, J. (1986) *Promised land: peasant rebellion in Chalatenango El Salvador.* LAB, London.

Pearse, A. (1975) *The Latin American peasant.* Frank Cass, London.

Pearse, A. (1980) *Seeds of plenty, seeds of want: social and economic implications of the green revolution.* Oxford University Press, Oxford.

Peattie, L. (1982) What is to be done with the 'informal sector'? A case study of shoe manufacturers in Colombia. In Safa, H. (1982) *Towards a political economy of urbanization in Third World countries.* Oxford University Press, Delhi.

Pepper, D. (1984) *The roots of modern environmentalism.* Croom Helm, London.

Perlman, J. (1976) *The myth of marginality: urban poverty and politics in Rio de Janeiro.* University of California Press, Berkeley and London.

Petras, J. (1969) *Political and social forces in Chilean development.* University of California Press, Berkeley and London.

Petras, J. (1981) Nicaragua: the transition to a new society. *Latin American Perspectives* 29, Vol. 8, No. 2.

Petras, J. (1992) *Latin America in the time of cholera: electoral politics, market economics and permanent crisis.* Routledge, New York and London.

Petras, J., Cooke, T. (1973) Dependency and the industrial bourgeoisie: attitudes of Argentinian executives towards foreign economic investments and US policy. In Petras, J. (ed.) *Latin America: from dependence to revolution.* Wiley, Chichester.

Petras, J., La Porte, R. (1971) *Cultivating revolution: the United States and agrarian reform in Latin America.* Vintage Books, New York.

Philip, G. (1985) *The military in South American politics.* Croom Helm, Beckenham, Kent.

Pitt-Rivers, J. (1977) *The fate of shechem or the politics of sex: essays in the anthropology of the Mediterranean.* Cambridge University Press, Cambridge.

Poblete, R. (1970) The church in Latin America: a historical survey. In Landsburger, H. (ed.) *The church and social change in Latin America.* University of Notre Dame Press, London.

Poole, D., Renique, G. (1992) *Peru, time of fear.* LAB, London.

Portes, A. (1970) Los grupos urbanos marginados: nuevo intento de explicación. *Aportes* 18.

Portes, A. (1985) Latin American class structures: their composition and change during the last decades. *Latin American Research Review* 20(3).

Portes, A. (1989) Latin American urbanization during the years of the crisis. *Latin American Research Review* **24**(3).

Poulantzas, N. (1975) *Political power and social classes.* New Left Books, London.

Prance, G. (1990) Fruits of the rain forest. *New Scientist* 13/11/90: 42–4.

PREALC (1981) Dinámica del subempleo en América latina. *Estudios informes de la CEPAL*, series 10, ECLAC, Santiago.

Preston, D. (1987) (2nd edn) *Latin American development, geographical perspectives.* Longman, Harlow.

Puig, J. (1990) Larga marcha por la vida. *Revista Nuestro Ambientes* 15(3).

Quadri, G. (1990) Una breve crónica de ecologismo en México. *Revista Ciencias* **4**.

Quijano, A. (1971) *Nationalism and capitalism in Peru: a study of neo-imperialism.* Monthly Review Press, New York and London.

Quijano, A. (1974) The marginal pole of the economy, and the marginalised labour force. *Economy and Society* **3**(4).

Radcliffe, S., Westwood, S. (1993) *VIVA! Women and popular protest in Latin America.* Routledge, London and New York.

Rama, G. (1983) Education in Latin America: exclusion or participation. *CEPAL Review* 21.

Ransom, D. (1992) Green justice. *The New Internationalist*, April: 4–7.

Rattansi, A., Boyne, A. (1990) *Postmodernism and society.* Macmillan, London.

Ratinoff, L. (1967) The new urban groups: the middle classes. In **Lipset, S. and Solari, A.** (eds) *Elites in Latin America.* Oxford University Press, London.

Redclift, M. (1984a) 'Urban bias' and rural poverty: a Latin American perspective. *Journal of Development Studies* **20**(3).

Redclift, M. (1984b) *Development and the environmental crisis: red or green alternatives?* Methuen, London.

Redclift, M. (1987) *Sustainable development: exploring the contradictions.* Methuen, London and New York.

Redfield, R. (1941) *The folk culture of Yucatan.* University of Chicago Press, Chicago.

Remmer, K. (1991) *Military rule in Latin America.* Westview Press, Boulder.

Ritchie, M. (1992) Free trade versus sustainable agriculture: the implications of NAFTA. *The Ecologist* 22/5: 221–7.

Robbins, J. (1987) *Diet for a new America.* Stillpoint Publishing, Canada.

Roberts, B. (1978) *Cities of peasants: the political economy of urbanisation in the Third World.* Edward Arnold, London.

Rodda, A. (1991) *Women and the environment.* Zed Books, London.

Rosenthal, G. (1989) Latin American and Caribbean development in the 1980s and the outlook for the future. *CEPAL Review* **39**, ECLAC, Chile.

Latin American Society

Rostow, W. (1960) *The stages of economic growth: a non-communist manifesto*. Cambridge University Press, Cambridge.

Rothstein, F. (1979) The class basis of patron–client relations. *Latin American Perspectives* 21, Vol. 6, No. 2.

Roxborough, I. (1979) *Theories of underdevelopment*. Macmillan, London.

Roxborough, I. (1984) *Unions and politics in Mexico: the case of the automobile industry*. Cambridge University Press, Cambridge.

Rubbo, A. (1975) The spread of capitalism in rural Colombia: effects on poor women. In Reiter, R. (ed.) *Toward an anthropology of women*. Monthly Review Press, New York and London.

Ruiz-Perez, S. (1979) Begging as an occupation in San Cristobal de las Casas, Mexico. In Bromley, R. and Gerry, C. (eds) *Casual work and poverty in Third World cities*. John Wiley, Chichester.

Safa, H. (ed.) (1977) The changing class composition of the female labour force in Latin America. *Latin American Perspectives* 15, Vol. 4, No. 4.

Safa, H. (1982) *Towards a political economy of urbanization in Third World countries*. Oxford University Press, Delhi.

Saffioti, H. (1978) *Women in class society* (translated from Portuguese by M. Vale). Monthly Review Press, London.

Said, E. (1978) *Orientalism*. Pantheon Books, New York.

Sarduy, P., Stubbs, J. (eds) (1993) *Afro Cuba: an anthology of Cuban writing on race, politics and culture*. LAB, London.

Schirmer, J. (1993) The seeking of truth and the gendering of consciousness: The COMADRES and the CONAVIGUA widows of Guatemala. In Radcliffe, S. and Westwood, S. (1993) *VIVA! Women and popular protest in Latin America*. Routledge, London and New York.

Schmink, M. (1977) Dependent development and the division of labour by sex: Venezuela. *Latin American Perspectives* 12 and 13, Vol. 4, Nos 1 and 2.

Schmitz, H. (1982) *Manufacturing in the backyard, with case studies on accumulation and employment in small scale Brazilian industry*. F. Pinter, London.

Selcher, W. (1981) Brazil in the world: multipolarity as seen by a peripheral ADC middle power. In Ferris, E. and Lincoln, J. (eds) *Latin American foreign policies: global and regional dimensions*. Westview Press, Boulder.

Seligson, M. (1980) *Peasants in Costa Rica and the development of agrarian capitalism*. University of Wisconsin Press, Madison and London.

Seligson, M. (ed.) (1984) *The gap between the rich and the poor*. Westview Press, Boulder.

Shanin, T. (1971) Peasantry as a political factor. In Shanin, T. (ed.) *Peasants and peasant societies*. Penguin, Harmondsworth.

Bibliography

Shanin, T. (1987) (2nd edn) *Peasants and peasant society*. Basil
Blackwell, Oxford.

Shiva, V. (1988) *Women's ecology and development*, Zed Books, London.

Simmons, A., Diaz-Briquets, S., Laguian, A. (1977) *Social change
and internal migration: a review of research findings from Africa, Asia
and Latin America*. International Development Centre, Ottawa.

Simmons, P. (1992) Women and development – a threat to liberation.
The Ecologist 22/1: 16–21.

Singer, H., Ansari, J. (1988) (4th edn) *Rich and poor countries:
consequences of international economic disorder*. Allen & Unwin,
London.

Singer, P. (1982) Neighbourhood movements in Sao Paulo. In Safa, H.
(ed.) *Towards a political economy of urbanization in Third World
countries*. Oxford University Press, Delhi.

Skidmore, T., Smith, P. (1989) (2nd edn) *Modern Latin America*.
Oxford University Press, Oxford and New York.

Slater, D. (1985) *New social movements and the state in Latin America.
Latin American Studies* 29. CEDLA, Amsterdam.

Smith, B.H. (1982) *The church and politics in Chile: challenges to modern
catholicism*. Princeton University Press, Guildford, Surrey.

Spalding, H. (1977) *Organised labour in Latin America: historical case
studies of workers in independent societies*. Harper & Row, New York.

Statistical Abstract of Latin America (1993) UCLA, Los Angeles.

Stavenhagen, R. (1970) Classes, colonialism and acculturation. In
Horrowitz, I. (ed.) *Masses in Latin America*. Oxford University
Press, New York.

Stavenhagen, R.. (1974) The future of peasants in Mexico. In Institute
of Latin American Studies, *The rural society of Latin America today*.
Almquivist & Wiksell, Stockholm.

Stavenhagen, R. (1978) Capitalism and the peasantry in Mexico. *Latin
American Perspectives* 18, Vol. 5, No. 3.

Stein, S., Stein, B. (1970) *The colonial heritage of Latin America: essays
on economic dependence in perspective*. Oxford University Press, New
York.

Stepan, A. (1973) *Authoritarian Brazil: origins, policies and future*. Yale
University Press, New Haven and London.

Stephen, L. (1992) Women in Mexico's popular movements: survival
strategies against ecological and economic impoverishment. *Latin
American Perspectives* 72, Vol. 19, No. 1 (Winter): 73–96.

Streeten, P. (1981) *Development perspectives*. Macmillan, London.

Strickon, A., Greenfield, S. (1972) (eds) *Structure and process in Latin
America: patronage, clientage and power systems*. University of New
Mexico Press, Albuquerque.

Stubbs, J. (1989) *Cuba: the test of time*. LAB, London.

Sunkel, O. (1965) Change and frustration in Chile. In Veliz, V. (ed.)

265

Obstacles to change in Latin America. Oxford University Press, London.

Taylor-Dormond, M. (1991) The state and poverty in Costa Rica *CEPAL Review* 43.

Third World Guide 93–94 (1992) (TWG) Instituto del Tercer Mundo, Uruguay. Distributed by Oxfam.

Tironi, E., Lagos, R. (1991) The social actors and structural adjustment. *CEPAL Review* 44, ECLAC, Chile.

Todaro, M.P. (1992) (3rd edn) *Economics for a developing world.* Longman, London and New York.

Tokman, V. (1982) Unequal development and the absorption of labour. *CEPAL Review* 17.

Treece, D. (1987) *Bound in misery and iron: the impact of Grande Carajás programme on the Indians of Brazil.* A report from Survival International, London.

Treece, D. (1990) Indigenous people in Brazilian Amazonia and the expansion of the economic frontier. In Goodman, D. and Hall, A. (1990) *The future of Amazonia: destruction or sustainable development?* Macmillan, London.

Turner, J. (1965) Lima's barriada and corralones: suburbs versus slums. *Ekistics* 112.

Turner, J. (1970) Barriers and channels for housing development in modernizing countries. In Mangin, W. (ed.) *Peasants in cities: readings in the anthropology of urbanisation.* Houghton Mifflin Co., Boston.

Underwood, J. (1993) *African cultural frontiers in Central America.* Dissertation submitted for BA honours degree in Latin American studies, Portsmouth University.

United Nations (1992) *UN demographic year book.* Department of Economic and Social Development, Statistics Department. UN, New York.

UNICEF (1981) *New dimensions, fair conditions.* UN, Geneva.

Vallier, I. (1967) Religious elites: differentiations and developments in Roman Catholicism. In Lipset, S. and Solari, A. *Elites in Latin America.* Oxford University Press, London.

van den Berghe, P., Primov, G. (1977) *Inequality in the Peruvian Andes: class and ethnicity in Cuzco.* University of Missouri Press, London.

Vasquez de Miranda, G. (1977) Women's labour force participation in a developing society: the case of Brazil. In The Wellesley Editorial Committee (eds) *Women and national development: the complexities of change.* University of Chicago Press, Chicago and London.

Vickers, J. (1991) *Women and the world economic crisis.* Zed Books, London and New Jersey.

Wade, P. (1986) Patterns of race in Colombia. *Bulletin of Latin American Research* 5(2): 1–9.

Walby, S. (1992) Post-post-modernism? Theorizing social complexity. In **Barrett, M. and Phillips, A.** *Destabilizing theory: contemporary feminist debates.* Polity Press, Oxford.

Wallace, M. (1990) (3rd edn) *Black macho and the myth of the superwoman.* Verso, London and New York.

Wallerstein, I. (1974) *The modern world system, 1: Capitalist agriculture and the origins of the world economy in the sixteenth century.* Academic Press, New York and London.

Wallerstein, I. (1980) *The modern world system, 2: Mercantilism and the consolidation of the European world economy, 1600–1750,* Academic Press, New York and London.

Ward, P. (1990) *Mexico City: the production and reproduction of the urban environment.* Belhaven Press, London.

Warren, B. (1973) Imperialism and capitalist industrialisation. *New Left Review* 81.

Weber, M. (1965) *The protestant ethic and the spirit of capitalism.* Allen & Unwin, London.

Weffort, F. (1970) Notas sobre 'la teoria de la dependencia': teoría de clase o ideología nacional. *Revista Latinoamericano de Ciencia Política* 1(3), December.

Weisskoff and Figueroa (1976) Traversing the social pyramid. *Latin American Research Review* 11(2).

Wilkio, J. (ed.) (1993) *Statistical Analysis of Latin America (SALA),* 30(1), UCLA Centre Publications, Ca.

Wilson, F. (1991) *Sweaters: gender, class and workshop based industry in Mexico.* Macmillan, Basingstoke and London.

Winant, H. (1992) Rethinking race in Brazil. *Journal of Latin American Studies* 24(1): 173–92.

Wolf, E. (1969) *Peasant wars of the twentieth century.* Harper & Row, New York.

Wolf, E., Hansen, E. (1972) *The human condition in Latin America.* Oxford University Press, New York.

Wood, C., Carvalho, J. (1988) *The demography of inequality in Brazil.* Cambridge University Press, Cambridge.

Woodgate, G. (1991) Agroecological possibilities and organisational limits: some initial impressions from a Mexican case study. In **Goodman, D. and Redclift, M.** (1991) *Environment and development in Latin America: the politics of sustainability.* Manchester University Press, Manchester.

World Bank (1992) *World development report.* Oxford and New York.

World Economic Survey – Current trends and policies in the world economy. (1992) Department of Economic and Social Development, United Nations, New York.

World in Figures (1987) *The Economist,* London.

World Resources (WR) *A guide to the global environment* (1986) Published in collaboration with the United Nations Environment

Programme, Oxford University Press, New York and Oxford.
World Resources (WR) *A guide to the global environment* (1992–1993). Published in collaboration with the United Nations Environment Programme, Oxford University Press, New York and Oxford.
World Tables (1991) Johns Hopkins for the World Bank, Baltimore and London.
Wuyts, M., Mackintosh, M., Hewitt, T. (1992) *Development policy and public action.* Oxford University Press, Oxford.
Young, K. (1978) Changing economic roles of women in two rural Mexican communities. *Sociologia Ruralis* 18(2/3).

Subject index

agrarian reform, 33–4, 79, 128, 134, 136–41, 144, 146, 148, 149, 179, 220
agribusiness, 40, 42, 75, 132, 133, 135, 142, 148, 220
Aztec, 84

bureaucratic technical class, 178, 179, 180, 183, 186, 195, 199
bourgeoisie, 34, 40, 42, 183, 184, 185, 186, 199, 201, 204, 206, 207, 210
  bourgeoisie capitalism, 32
  bourgeoisie state, 183
  comprador bourgeoisie, 182
  national bourgeoisie, 182
  petty bourgeoisie, 146, 178, 181, 220
  state bourgeoisie, 182, 201

cholera, 53, 93, 94, 217
church, 3, 76, 83–8, 89, 93, 99, 104, 109, 110, 129, 130, 149
clientelism, 80, 104–6, 139, 147, 190, 191, 192, 206, 211, 217
compadrazgo, 103–5
cooperatives, 129, 137, 140, 159

debt (national), 4, 22, 24, 29, 45, 51, 52, 54, 60, 67, 68, 69–70, 71, 75, 94, 118, 142, 152, 196, 209, 210, 216, 217, 220, 222, 223, 224, 225
dependency theory, 13, 34–45, 46, 47, 50, 55, 56, 86, 94, 97, 98, 99, 143–4, 147, 153, 179, 182, 185, 204, 206, 216, 221, 225
dominant class, 105, 177, 178, 179, 180, 182, 183, 186, 195, 199, 207, 208, 210
dualism, 26, 33, 39, 41, 49, 54, 168, 206

ECLA, 35–6, 214
economic growth, 1–2, 4, 7, 8, 11–12, 14, 15, 17, 20, 25, 26, 27, 28, 33, 34, 39, 40, 44, 45, 51, 52, 57, 59–66, 68, 70, 81, 106, 136, 139, 145, 152, 190, 199, 201, 202, 211–14, 216, 217, 220–4
education, 1–2, 6, 10, 13–16, 17, 50, 53, 64, 74, 84, 85, 89–93, 98, 100, 111, 112, 117, 118, 120, 121, 126, 137, 143, 149, 153, 154, 157, 158, 163, 173, 184, 201, 202, 206, 211, 212, 215, 217, 222, 224

Place name index